The Neurobiological and Developmental Basis for Psychotherapeutic Intervention

THE LIBRARY OF CLINICAL PSYCHOANALYSIS

A Series of Books Edited By

Steven J. Ellman

This series is intended to show the depth, flexibility, and vigor of contemporary Freudian thought. It is by no means a series that can be equated with classical psychoanalysis or ego psychology, but rather is intended to show how contemporary Freudians are able to meaningfully integrate a variety of positions into the scaffolding of Freudian tenets. Goldman's and especially Bach's work show the richness and depth that can be achieved within such a framework.

Freud's Technique Papers: *A Contemporary Perspective*
Steven J. Ellman

In Search of the Real: *The Origins and Originality of D.W. Winnicott*
Dodi Goldman

In One's Bones: *The Clinical Genius of Winnicott*
Dodi Goldman, Editor

The Language of Perversion and the Language of Love
Sheldon Bach

Omnipotent Fantasies and the Vulnerable Self
Carolyn S. Ellman and Joseph Reppen, Editors

The Neurobiological and Developmental Basis for Psychotherapeutic Intervention
Michael Moskowitz, Catherine Monk, Carol Kaye, and Steven J. Ellman, Editors

Enactment: *A New Paradigm*
Steven J. Ellman and Michael Moskowitz, Editors

The Neurobiological and Developmental Basis for Psychotherapeutic Intervention

edited by
Michael Moskowitz, Ph.D.
Catherine Monk, Ph.D.
Carol Kaye, Ph.D.
Steven J. Ellman, Ph.D.

JASON ARONSON INC.
Northvale, New Jersey
London

This book was set in 12 pt. Galliard by FASTpages of Nanuet, New York and printed and bound by Book-mart Press, Inc. of North Bergen, New Jersey.

Library of Congress Cataloging-in-Publication Data

The neurobiological and developmental basis for psychotherapeutic
    intervention / edited by Michael Moskowitz . . . [et al.].
        p.   cm.
    Includes bibliographical references and index.
    ISBN 0-7657-0097-2 (alk. paper)
    1. Object relations (Psychoanalysis)—Congresses.
2. Psychoanalysis—Congresses.   3. Mother and infant—Congresses.
4. Self-control—Congresses.   5. Internalization—Congresses.
I. Moskowitz, Michael.
    [DNLM: 1. Self Concept—congresses.   2. Psychoanalytic Theory—
congresses.   3. Child Development—congresses.   4. Mother–Child
Relations—congresses.   5. Developmental Disabilities—etiology—
congresses.   6. Developmental Disabilities—psychology—congresses.
WM 460.5.E3 C641 1997]
RC455.4.O23C58   1997
616.89'17—dc21
DNLM/DLC
for Library of Congress   97–23572

Printed in the United States of America on acid-free paper. Jason Aronson Inc. offers books and cassettes. For information and catalog write to Jason Aronson Inc., 230 Livingston Street, Northvale, New Jersey 07647-1731. Or visit our website: http://www.aronson.com

Contents

Editors and Contributors

Beatrice Beebe, Ph.D., is an Associate Clinical Professor of Psychology in Psychiatry at Columbia University; a faculty member of the New York University Postdoctoral Program in Psychotherapy and Psychoanalysis; and a founding faculty member of the Institute for the Psychoanalytic Study of Subjectivity. She is in private practice in New York City.

Patricia Daniel is a Training and Supervising Analyst at the British Psycho-Analytical Society. She is affiliated with the Tavistock Clinic and works with children and adults.

Andrew B. Druck, Ph.D., is President-elect, Dean, Fellow (Training and Supervising Analyst), and faculty member at the Institute for Psychoanalytic Training and Research (IPTAR) in New York City. He is also Clinical Assistant Professor of Psychology in the New York University Postdoctoral Program in Psychotherapy and Psychoanalysis. Dr. Druck is the author of *Four Therapeutic Approaches to the Borderline Patient*.

Steven Ellman, Ph.D., is Co-Chair of the Independent Psychoanalytic Societies. He is Past President and Program Chair of the Institute for Psychoanalytic Training and Research (IPTAR) in New York City, Professor in the Doctoral Program in Clinical Psychology at the City University of New York, and faculty member and supervisor in the New York University Postdoctoral Program in Psychotherapy and Psychoanalysis.

James L. Fosshage, Ph.D., is Co-founder and Board Director of the National Institute for the Psychotherapies, New York City, Core Faculty member, Institute for the Psychoanalytic Study of Subjectivity, New York City, and Clinical Professor of Psychology, New York University Postdoctoral Program in Psychotherapy and Psychoanalysis. His most recent books (co-authored with Joseph Lichtenberg and Frank Lachmann) are *Self and Motivational Systems* and *The Clinical Exchange: Techniques Derived from the Standpoint of Self and Motivational Systems.*

Norbert Freedman, Ph.D., is Professor and Director of Psychology at the State University of New York, Health Science Center at Brooklyn; Adjunct Professor, New York University Postdoctoral Program in Psychoanalysis and Psychotherapy; Fellow and Past President of IPTAR (Institute for Psychoanalytic Training and Research); and Co-chair, Section on Investigative Psychoanalysis, IPTAR. His current work centers on the psychoanalytic theory of symbolization and desymbolization in clinical communication.

Carol Kaye, Ph.D., is former Clinical Professor of Psychology at New York University and Teachers' College, Columbia University, where she taught clinical child psychology. She is a member of the faculty and a Training and Supervising Analyst at the Institute for Psychoanalytic Training and Research (IPTAR). Dr. Kaye is in the private practice of adult psychotherapy and psychoanalysis in New York City.

Frank M. Lachmann, Ph.D., is a founding faculty member of the Institute for the Psychoanalytic Study of Subjectivity; a Training and Supervising Analyst at the Postgraduate Center for Mental Health; and a faculty member of the New York University Postdoctoral Program in Psychotherapy and Psychoanalysis. He has co-authored two books with Joseph Lichtenberg and James Fosshage: *Self and Motivational Systems* and *The Clinical Exchange: Techniques Derived from Self and Motivational Systems.*

Joseph D. Lichtenberg, M.D., is Editor-in-Chief of *Psychoanalytic Inquiry*, Director of the Institute for Contemporary Psychotherapy, and Clinical Professor of Psychiatry at Georgetown University. His most recent books (co-authored with Frank Lachmann and James Fosshage) are *Self and Motivational Systems* and *The Clinical Exchange: Techniques Derived from Self and Motivational Systems.*

Catherine Monk, Ph.D., is a research fellow in the Psychobiological Sciences Training Program, Columbia University College of Physicians and Surgeons and New York State

Psychiatric Institute. She is a candidate in psychoanalysis at the Institute for Psychoanalytic Training and Research.

Michael Moskowitz, Ph.D., is a publisher at Jason Aronson Inc. and an Adjunct Associate Professor in the City University of New York Clinical Psychology Program. He is co-editor (with RoseMarie Pérez Foster and Rafael Art. Javier) of *Reaching Across Boundaries of Culture and Class: Widening the Scope of Psychotherapy.* Dr. Moskowitz is a member of the Institute for Psychoanalytic Training and Research (IPTAR) and in private practice in New York City.

Julia L. Prillaman is a doctoral candidate in clinical psychology at Seton Hall University. Her current interests include psychodynamic treatment outcome research and the clinical application of object relations theory to adults and children.

Allan N. Schore, Ph.D., is Assistant Clinical Professor of Psychiatry and Biobehavioral Sciences, University of California at Los Angeles Medical School, and is a faculty member at the Institute of Contemporary Psychoanalysis in Los Angeles. Dr. Schore has authored an interdisciplinary volume, *Affect Regulation and the Regulation of the Self,* and is currently writing a clinical sequel, *Affect Regulation and the Repair of the Self.* He continues to publish articles and chapters across a spectrum of disciplines that are interested in the effects of early experiences on the developing brain and in integrative models of normal and abnormal socioemotional development—psychoanalysis, developmental psychopathology, infant research, neuroscience, and affect theory.

Arnold Wilson, Ph.D., is Professor of Psychology and Director of Clinical Training at Seton Hall University, and a faculty member at the Columbia University Center for Psychoanalytic Training and Research. Dr. Wilson is series co-editor, with Steven Ellman, of *Contemporary Issues in Psychoanalysis*, and co-editor, with John Gedo, of *Hierarchical Concepts in Psychoanalysis*.

Introduction*

Norbert Freedman and Michael Moskowitz

This book addresses the relationship between early experience and later difficulties in life. That there is such a relationship has been an accepted axiom of most modes of therapy since the invention of psychoanalysis as an iteration and amplification of the folk wisdom, "as the twig is bent, so grows the tree." An accepted premise in academic psychology from its beginnings, Thorndike wrote in 1905, "Every event of a man's [sic] mental life is written indelibly in the brain's archives, to be counted for or against him" (cited in Kagan 1996, p. 901). Psychoanalytic developmental theory has always been devoted to examining the links between early development and later life. More recently, perhaps starting with Loewald's (1960) seminal paper, the therapeutic process has been increasingly viewed as a venue for redoing distorted development or restarting arrested development.

But this accepted wisdom linking infancy and adult life has been challenged by many and arguably held either unproven or so vague as to be of no useful clinical value. The narrative or hermeneutic turn in psychoanalysis, from which we now appear to be recovering, can in part be

* *Editor's note:* This book is an outgrowth of Perspectives on Early Development, an IPTAR conference held in March 1996 and sponsored in part by Jason Aronson Inc. We wish to thank both IPTAR and Jason Aronson Inc. for their support and encouragement.

understood as stemming from a seeming inability to reliably link specific adult character traits or behavior to early experience. Academic psychologists currently responding to the inroads made by attachment research (which is re-opening the doors to a psychoanalytic academic presence) have raised questions about the predictive power of early experience. In a recent essay Jerome Kagan (1996) compares the thesis of the predictive power of early experience in general and the importance of mother–infant attachment, in particular to religious and superstitious beliefs. Michael Lewis (1997) attacks not only attachment research but most prevailing models of child development in arguing that early experience neither determines who we are nor limits what we become.

While some analysts (e.g., Schafer 1976) have questioned the validity of inferring preverbal experience from psychoanalytic data produced in putatively regressed states, most clinicians still remain convinced of the importance of early experience and the mnemonic and reparative power of the psychoanalytic setting. Yet the vagaries of the clinical situation, the variety of interpretations possible in any given situation, the disparate emphases of different theoretical perspectives (all in the context of an increasingly antagonistic antipsychoanalytic public environment) have led many to question the possibility of linking the clinical situation to accessible and replicable scientific studies of early development.

Into this intellectual climate emerged Allan Schore. With the publication of *Affect Regulation and the Origins of the*

Self: The Neurobiology of Emotional Development (1994), Schore provided a bridge between the academic developmental, neurobiological, and psychoanalytic worlds in his presentation of a comprehensive and empirically grounded theory of the development of the self and the ways in which the psychoanalytic situation provides a setting for the repair or remediation of distorted development. Digesting and integrating volumes of scientific literature (there are over 2300 references) Schore states unequivocally:

> The beginnings of living systems indelibly set the stage for every aspect of an organism's internal and external functioning throughout the lifespan. . . . Genetic systems that program the evolution of biological and psychological structures continue to be activated at very high rates over the stages of infancy, and this process is significantly influenced by factors in the postnatal environment. . . . The child's first relationship, the one with the mother, acts as a template, as it permanently molds the individual's capacities to enter into all later emotional relationships. [p. 3]

Schore amply illustrates how early interactions between mother and child shape the child's regulatory processes *at the level of the brain*, and that the social shaping of the brain has implications for a person's functioning throughout life.

With his emphasis on the origins of the self, Schore speaks to the concerns of psychoanalysts. The concept of the common origin of the self was ushered into the modern era by William James some one hundred years ago when he noted that from the buzzing, blooming confusion of infancy there emerges the monumental division of the mind between that which is ME and that which is NOT-ME. This idea has undergone many transformations, whether it is ego as doer, self as object, or self in potential space. It is Schore's work that has convincingly shown us that these essentially hidden internal processes are now amenable to new quantitative and qualitative observations. Moreover, by showing the continuity between self-regulation in infancy and self-regulation in adults in the clinical setting, Schore provides us with the underpinnings of the essential efficacy of our psychoanalytic enterprise, so crucial in our current climate of bombardment.

By emphasizing the neurophysiological aspects of development, Schore speaks to the knowledge base of our discipline. Russel Brain (1951) writes, "If we look at a tapestry closely, view it with a magnifying glass, we see the threads, but from a distance it is composed of patterns; the threads are the brain, the patterns are the mind" (p. 30). We are now beginning to see how the threads and the patterns, the mind and the brain, interact to form the whole person. By showing the convergence of critical interactive experiences in early development as they impinge upon and interact with the neurobiological development of brain structures, Schore gives us a vision that frees us from reductionism. Brain structures, in our current climate of biological psychiatry, have been assigned a primary deterministic and causal role, and hence the psyche is given a secondary role as epiphenomenon. Schore shows us that development at all levels, including in the analytic situation, can only be understood in terms of structure–function relationships.

By focusing on affect, explicit or implied, shame or mortification, overt or throttled, Schore speaks to the immediacy of the analytic situation as it affects central organizing processes. Affects are viewed as experience-dependent forces in the development of brain structures. Schore shows how affect-driven phenomena become the motives that determine human attachment, and indeed the organization of the self. To make a somewhat whimsical analogy, we might call our current sphere of inquiry the Post-Cartesian Era. We were all taught in our introductory philosophy classes that centuries ago Descartes told us about the origin of the self: *cogito ergo sum*: I think, therefore I am. Today we may say, *sento ergo sum*: I feel, therefore I am. This is the implicit vision of Dr. Schore's contribution.

In this volume Schore summarizes essential aspects of his integrative theory and elaborates on some clinically relevant implications. He shows how "the data that is being generated by the current explosion in infant research is not only giving us a more detailed model of human development, it is also being rapidly absorbed into clinical models where it is radically altering the central concepts of psychoanalysis and psychiatry" (p. 3). Among a few of the many often startling and provocative conclusions that Schore offers are:

- Regulatory transactions underlie the formation of attachment bonds between infant and caregiver.
- The concept of symbiosis is solidly grounded in research and should be retained.
- The core of the self is nonverbal and unconscious and lies in patterns of affect regulation.

- The concept of drive as a phenomenon at the border between the psychic and the somatic is needed and belongs back in psychoanalysis.
- A deeper examination of the interactive affect-trans-acting aspects of the nonverbal, unconscious transfer-ence–countertransference relationship is central to understanding therapeutic action.
- Yet, putting nonverbal affect states into words (interpretations) is an important achievement in self-regulation.

The remaining chapters in this volume speak to many of the issues that Schore presents. In Chapter 2 Ellman and Monk discuss Schore's contribution. While agreeing in general with Schore's synthesis of emotional and neurological development, they suggest the need to incorporate earlier mother–child experiences into our conceptualization of the development of emotion regulation. They argue that Schore's entry point in the mother–child story, seemingly after three months, with a focus on the interactions of an older child, leads to an overemphasis on the visual domain of communication as well as inaccuracies in Schore's meshing of psychoanalytic theory with neurobiological and social-emotional models of development. Further, Ellman and Monk show that even though Freudian theory must be updated to accommodate the new perspectives on mind-brain relationships presented in Schore's work, these additions and changes fit within Freud's essential postulates.

In Chapter 3, pioneering infant researchers Lachmann and Beebe offer a clinical case discussion focusing on the analyst's and patient's mutually regulated nonverbal interactions of affect, mood, arousal, and rhythm. They offer this approach neither as a new technique nor as a suggestion for diverting one's attention from dynamic formulations. Instead, they "reverse the figure-ground perspective" by putting the "mutually regulated nonverbal exchanges into the foreground and the dynamic conceptualization into the background" (p. 96).

Coming from different theoretical traditions, the next three chapters continue to address the theme of the relationship of early experience to the clinical psychoanalytic situation. From an intersubjective perspective, Fosshage and Lichtenberg, in Chapter 4, present a detailed clinical examination of a transference–countertransference interaction. They view transference not as a displacement from earlier relationships but as an interactive phenomenon that is variably affected by both patient and analyst. Even so, transference is not a wholly new creation but is built on schema that are reactivated in the new therapeutic situation.

From a Kleinian viewpoint, Daniel, in Chapter 5, using both infant–mother observation and adult analytic case material, looks to early development as the period for shaping unconscious dynamic conflicts, and demonstrates the role of internal conflict in transforming one's experience of the object and of reality itself. She shows how appropriate interpretations of unconscious conflict lead to an internalization of the analyst and help her patient move from a primitive paranoid-schizoid position to a developmentally more advanced depressive position.

In Chapter 6, from a contemporary Freudian perspective, Wilson and Prillaman show the effect of early development on later functioning by focusing on what they term *disorders of internalization* that affect patients' level of functioning as well as their ability to take in what the analyst has to offer. Drawing on research inside and outside the clinical domain, Wilson and Prillaman see problems in achieving self-regulation as interfering with internalization capacities and detail how treatment must then focus on the internalization disorders from inside the self regulatory system but also outside the transference.

In our concluding Chapter 7, Druck offers a discussion and synthesis. Focusing on the contributions of Daniel, Fosshage and Lichtenberg, and Wilson and Prillaman, he suggests that this theoretical panoply of contemporary Freudian (Wilson), Kleinian (Daniel), and intersubjective (Fosshage and Lichtenberg) perspectives all have a common ground in acknowledging that successful treatment addresses structural deficits manifested in problems of affect regulation and understood on the level of part-object (e.g., selfobject) relationships as well as intrapsychic and interpersonal conflicts that more often track relationships between people experienced as whole and separate others. But Druck offers no easy peace treaty. He points out significant differences among these perspectives, noting, for example, that the kinds of interpretations suggested by Daniel are precisely the wrong thing to do according to Fosshage and Lichtenberg because they will make the patient feel misunderstood.

Yet in listening to Daniel, it is hard to believe that her patients often feel misunderstood. Like the best analysts of all eras she says what she feels her patients can take in. This, however, is not an explicit aspect of her theory of technique. What the other contributors to this volume make explicit, and what is independently corroborated by the many researchers whose work Schore has brought together, is that the most correct understandings (and there may be several correct understandings) can be made use of by the patient only if the analyst is attuned to the patient's state at the time the interpretation is offered.

References:

Brain, R. (1951). *Mind, Perception and Science*. Oxford: Blackwell Scientific Publications.

Kagan, J. (1996). Three pleasing ideas. *American Psychologist* 51:901–908.

Lewis, M. (1997). *Altering Fate: Why the Past Does Not Predict the Future*. New York: Guilford.

Loewald, H. (1960). On the therapeutic action of psychoanalysis. *International Journal of Psycho-Analysis* 41:16–33.

Schafer, R. (1976). *A New Language for Psychoanalysis*. New Haven: Yale University Press.

Schore, A. (1994). *Affect Regulation and the Origins of the Self: The Neurobiology of Emotional Development*. Hillsdale, NJ: Lawrence Erlbaum.

1

Interdisciplinary Developmental Research as a Source of Clinical Models

Allan N. Schore

After remaining almost unchanged for most of psycho-analysis's first century, the central core of Freud's model of the mind is now undergoing a rapid and substantial transformation. The scaffolding of clinical psychoanalysis is supported by underlying theoretical conceptions of psychic development and structure, and it is these basic concepts that are now being reformulated. The elucidation of these meta-psychological underpinnings is doing more than strengthening clinical models—it is enriching the intellectual climate of our discipline. More than anyone could have predicted, observational and experimental research of infants interacting with their mothers has turned out to be the most fertile source for the generation of heuristic hypotheses about not only early development but also psychic dynamics. Indeed, a deeper understanding of the fundamental processes that drive development, of why early experience influences the organization of psychic structure and how this stucture comes to mediate emergent psychological functioning, of the origins of the human mind, is now within sight.

The question of why the early events of life have such an inordinate influence on literally everything that follows is one of the fundamental problems of not only psychoanalysis but of all science. How do early experiences, especially affective experiences with other humans, induce and organize the patterns of structural growth that result in the expanding functional capacities of a developing individual? A spectrum of disciplines—from developmental biology and neurochemistry through developmental psychology and psychoanalysis—

share the common principle that the beginnings of living systems indelibly set the stage for every aspect of an organism's internal and external functioning throughout the lifespan. A developmental theory, a conception of the genesis of living systems, a model of self organization, is found at the base of each and every domain of theoretical and clinical science. The data that is being generated by the current explosion in infant research is not only giving us a more detailed model of human development, it is also being rapidly absorbed into clinical models where it is radically altering the central concepts of psychoanalysis and psychiatry. At present all major theoreticians are placing developmental concepts at the foundation of their clinical models.

Using an expanding arsenal of different methodologies and studying different levels of analysis, multidisciplinary investigators are now inquiring into the incipient interactions the infant has with the most important object in the early environment—the primary caregiver. It is now very clear that *this dialectic with the social environment is mediated by transactions of affect, and that this emotional communication is nonverbal.* Human development cannot be understood apart from this affect-transacting relationship. Furthermore, these early social events are imprinted into the biological structures that are maturing during the brain growth spurt that occurs in the first two years of human life, and therefore have far-reaching and long enduring effects (see Figure 1–1). In the November issue of the *American Journal of Psychiatry*, Eisenberg (1995) presents an article, "The *Social* Construction of the Human Brain" (italics added). It is now well

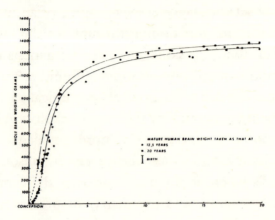

Figure 1-1: Growth of whole brain compared when mature weight is taken at 12.5 years and at 20 years. Note the accelerated growth in the first two years (from Himwich 1975).

established that the accelerated growth of brain structure during critical periods of infancy is experience-dependent and influenced by social forces. But neurobiology is unclear as to the nature of these "social forces."

In fact, psychoanalysis has much to say about the social forces that influence the organization of developing psychic structure. The period of early-forming object relations so intensely studied by psychoanalysis exactly overlaps the period of the brain growth spurt. The most important revisions of Freud's model in the latter half of this century have occurred in developmental psychoanalysis—Fairbairn (proclaiming that infants are not solely instinctually driven but object-seeking), Klein (exploring very early developmental events and primitive cognitive mechanisms), Winnicott (concluding that an infant cannot be understood apart from its interaction with the mother), Bowlby (applying then current biology to an understanding of infant–mother attachment), Mahler (introducing observational research in psychoanalysis), Stern (focusing on interactive attunement mechanisms in the relationship), and Emde (emphasizing that the child's environment is the relationship with the primary caregiver).

There are few visual symbols in psychoanalysis. The symbol of classical psychoanalysis is a photograph of Freud's face (Figure 1–2), an icon of a monad, a single unit; an adult, conscious, reflective mind attempting to understand the realm of the dynamic unconscious that forms in early childhood; a man's face gazing inward; a representation of a paternal-oedipal psychology.

But there is one other visual symbol of developmental psychoanalysis that is familiar to us—the icon from the cover of Stern's (1985) book *The Interpersonal World of the Infant* (Figure 1–3). Here we have a symbol of a dyad: two interlocking units; gazes between two faces, one of an adult female, the other of an infant; a representation of a maternal-preoedipal psychology. This is, of course, Mary Cassatt's *Baby's First Caress*, painted in 1890. The two figures form a compact and unified group; thus the structure expresses a close maternal-infant relationship. The visual image conveys their Winnicottian intense engagement, a Bowlbian image of bonding. How does the artist convey this? Yes, the curve of their arms suggests a loop, the baby touches the mother's face, the mother's hand is on the baby's foot—a closed system of linked bodies. But more than this, the heads are placed together; what unites them is the meeting of *direct glances between their eyes.*

Figure 1–2: Freud, age 53

My book, *Affect Regulation and the Origin of the Self*, enlarges and concentrates on this focus. In this volume, I have integrated new insights about interactive, relational pro-

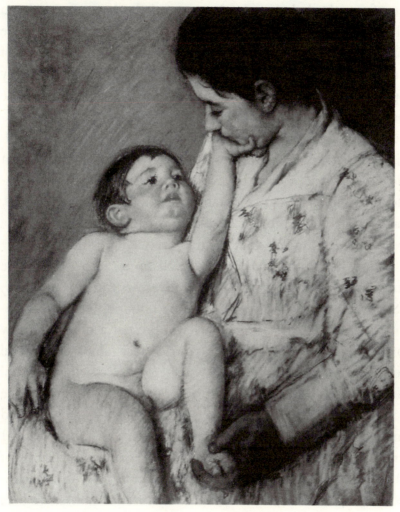

Figure 1–3: *Baby's First Caress* by Mary Cassatt

cesses from developmental and clinical psychoanalysis, current ideas about the origins of social functioning from the developmental sciences, recent data on emotional phenomena from the behavioral sciences, and the latest research on the limbic structures that support these functions from the brain sciences, in order to generate an overarching model of emotional development (Schore 1994). A multidisciplinary approach is especially relevant to the investigations of the psychobiological mechanisms that underlie affective processes. In contrast to reductionism, my work presents a multilevel perspective that is guided by the tenet that development can only be explained by studying it simultaneously along several separate but interrelated dimensions, ranging from the biological level of organization through the psychological, social, and cultural levels.

Drawing upon my book, as well as on an article in the interdisciplinary journal *Development and Psychopathology* (Schore 1996) and on an upcoming volume, *Affect Regulation and the Repair of the Self,* I will present a brief overview of recent studies from a spectrum of disciplines that reveal more detailed and precise knowledge of early socioemotional development. This information about the ontogeny of the mind cannot be apprehended clinically, yet it is relevant and indeed essential to psychoanalysis. As you will soon notice, I am arguing that the appearance of the adaptive functions of the developing mind cannot be understood without also addressing the problem of the maturation of structures responsible for these functions. Changes in the child's behavior or in the child's internal world can only be understood in terms of the appearance of a more complex structure that performs emergent functions. In describing the process of psychological structure formation, Rapaport declared, "We must establish how processes turn into structures, how a structure once formed changes, and how it gives rise to and influences processes" (1960, pp. 98–99). Freud's structural theory must not be abandoned but updated in terms of what we now know about brain-mind relationships. Psychoanalysis must now come to terms with structure, and its models of psychic structure should not be reduced to neurobiology, but should be compatible with what is now known about structure as it exists in nature.

Toward that end, I will begin by offering a multilevel perspective of the structure-function relationships of an event central to human emotional development—the interactive creation of an attachment bond of affective communication between the primary caregiver and the infant. In the course of this I will outline psychobiological models of the mirroring process, symbiosis, and selfobject phenomena. I will then describe how these very same affect-transacting experiences specifically shape the maturation of specific structural connections within the brain that come to mediate both the interpersonal and intrapsychic aspects of all future socioemotional functions. Of particular importance is the organization of a hierarchical regulatory system in the prefrontal areas of the right hemisphere. Finally, I will begin to sketch out a few of the important implications of these results for contemporary theoretical and clinical psychoanalysis, which Cooper (1987) asserts is "anchored in its scientific base in developmental psychology and in the biology of attachment and affects" (p. 83).

Affective Transmissions in Mutual Gaze Transactions

Although much has been written about cognitive development in infancy, until very recently few studies have been done on emotional and social ontogeny. This development is closely tied into the maturation of sensory systems, especially visual systems. In fact, over the first year of life *visual experiences play a paramount role in social and emotional development* (Blank 1975, Fraiberg and Freedman 1964, Hobson 1993, Keeler 1958, Nagera and Colonna 1965, Preisler 1995, Wright 1991). In particular, the mother's emotionally expressive face is, by far, the most potent visual stimulus in the infant's environment, and the child's intense interest in her face, especially in her eyes, leads him to track it in space and to engage in periods of intense mutual gaze. The infant's gaze, in turn, reliably evokes the mother's gaze, thereby acting as a potent interpersonal channel for the transmission of reciprocal mutual influences. These sustained face-to-face transactions are quite common and can be of very long duration. They mediate what Spitz (1958) calls "the dialogue between mother and child." In fact, gaze represents the most intense form of interpersonal communication, and the perception of facial expressions is known to be the most salient channel of nonverbal communication.

Congruent with these findings, Kohut (1971) concludes: "The most significant relevant basic interactions between mother and child usually lie in the visual area: The child's bodily display is responded to by the gleam in the mother's eye" (p. 117). There is now evidence that the mother's gleam is more than a metaphor. By 2 to 3 months, a time of increasing myelination of the visual areas of the infant's occipital cortex, the mother's eyes become a focus of her infant's attention, especially her pupils. Studies by Hess (1975) show that a woman's eyes (and those of a man with children) dilate in response to the image of a baby, a response associated with the positive emotions of pleasure and interest. Furthermore, an infant will smile in response to enlarged pupils. Even more intriguingly, viewing enlarged pupils rapidly elicits dilated pupils in the baby, and dilated pupils are known to release caregiver behavior. The pupil of the eye thus acts as an interpersonal nonverbal communication device, and these rapid communications occur at unconscious levels. Hess concludes, ". . . the fact that babies have large pupils, or respond with enlarging pupils in adult–infant interaction would in general assure at least a minimal degree of the infant–adult interaction that is necessary for the mental and emotional development of the child" (p. 106).

Mutual gaze interactions increase over the second and third quarter of the first year, and since they occur within the "split second world" of the mother and infant (Stern 1977) they are therefore not easily visible. This dialogue is best studied by a frame-by-frame analysis of film, and in such work Beebe and Lachmann (1988) observe synchronous rapid movements and fast changes in affective expressions within the dyad (see Figure 1–4). This affective mirroring is accomplished by a moment-by-moment matching of affective direction in which both partners increase their degree of engagement and facially expressed positive affect together.

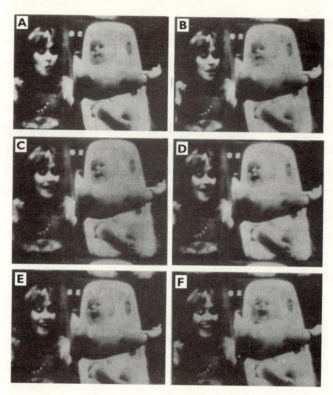

Figure 1–4: Photographic illustrations of a mirroring sequence. Mother and infant are seated face to face, and are looking at each other. At point A, mother shows a "kiss-face," and infant's lips are partially drawn in, resulting in a tight, sober-faced expression. At point B, .54 seconds later, mother's mouth has widened into a slightly positive expression, and infant's face has relaxed with a hint of widening in the mouth and a slightly positive expression. At point C, .79 seconds later, both mother and infant show a slight smile. At point D, .46 seconds later, both mother and infant further widen and open their smiles. Again at points E, .46 seconds later, and F, .58 seconds later, both mother and infant further increase their smile display. Points E and F illustrate the infant "gape smile." At point F the infant has shifted the orientation of his head further to the left and upward, which heightens the evocativeness of the gape smile (from Beebe and Lachmann 1988).

The fact that the coordination of responses is so rapid suggests the existence of a bond of unconscious communication.

This microregulation continues, as, soon after the "heightened affective moment" of an intensely joyful full gape smile the baby will avert his gaze in order to regulate the potentially disorganizing effect of this intensifying emotion (Field and Fogel 1982, Figure 1–5). In order to maintain the positive emotion the attuned mother takes her cue and backs off to reduce her stimulation. She then waits for the baby's signals for reengagement. Importantly, not only the tempo of their engagement but also that of their disengagement and reengagement is coordinated. In this process of contingent responsivity the more the mother tunes her activity level to the infant

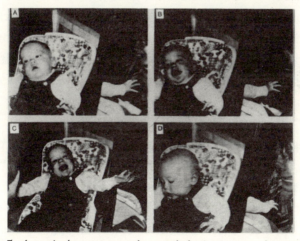

Figure 1–5: A typical sequence observed during attuned interactions of normal infants and their mothers: (a) the infant looks at the mother and the mother shows an exaggerated facial expression (mock surprise); (b) the infant and mother smile; (c) the infant laughs, the mother "relaxes" her smile; and (d) the infant looks away and the mother ceases smiling and watches her infant (from Field and Fogel 1982).

during periods of social engagement, the more she allows him to recover quietly in periods of disengagement, and the more she attends to his reinitiating cues for reengagement, the more synchronized their interaction. Facial mirroring thus illustrates interactions organized by ongoing regulations, and experiences of mutually attuned synchronized interactions are fundamental to the ongoing affective development of the infant (Feldman et al. 1996, Feldman and Greenbaum 1997).

These mirroring exchanges generate much more than overt facial changes in the dyad; they represent a transformation of inner events. Beebe and Lachmann (1988) assert that as the mother and infant match each other's temporal and affective patterns, each recreates *an inner psychophysiological state similar to the partner's.* In synchronized gaze the dyad creates a mutual regulatory system of arousal (Stern 1983) in which they both experience a state transition as they move together from a state of neutral affect and arousal to one of heightened positive emotion and high arousal. The mother's face, the child's emotional or biological mirror, has been described as reflecting back her baby's "aliveness" in a "positively amplifying circuit mutually affirming both partners" (Wright 1991, p. 12). Stern (1985) refers to a particular maternal social behavior that can "blast the infant into the next orbit of positive excitation" and generate "vitality affects" (p. 197).

In order to enter into this communication, the mother must be psychobiologically attuned not so much to the child's overt behavior as to the reflections of his internal state. She also must monitor her own internal signals and differentiate her own affective state, as well as modulate nonoptimal high levels of stimulation that would induce supra-heightened levels of arousal in the infant. The burgeoning capacity of the infant to experience increasing levels of accelerating, rewarding affects (enjoyment-joy and interest-excitement; Tomkins 1962) is thus at this stage externally regulated by the psychobiologically attuned mother, and depends upon her capacity to engage in an interactive emotion-communicating mechanism that generates these in herself and her child.

We now know that the caregiver is not always attuned; indeed, developmental research shows frequent moments of misattunement in the dyad, ruptures of the attachment bond. In fact, over the course of time, especially as the baby becomes a mobile toddler, she shifts from a caregiver to a socialization agent, and in doing so she continues to use visual channels for emotional communication. Reciprocal gaze, in addition to transmitting attunement, can also act to transmit misattunement, as in shame experiences. The misattunement in shame, as in other negative affects, represents a regulatory failure, and is phenomenologically experienced as a discontinuity in what Winnicott (1958) calls the child's need for "going-on-being."

Prolonged negative states are too toxic for infants to sustain for very long, and although they possess some capacity to modulate low intensity negative affect states, these states continue to escalate in intensity, frequency, and duration. How long the child remains in states of intense negative affect is an important factor in the etiology of a predisposition to psychopathology. Active parental participation in state regulation is critical to enabling the child to shift from the negative affective states of hyperaroused distress or hypoaroused deflation to a reestablished state of positive affect. In early development an adult provides much of the necessary modulation of infant states, especially after a state disruption and across a transition between states, and this allows for the development of self-regulation. Again, the key to this is the caregiver's capacity to monitor and regulate her own affect, especially negative affect.

In this essential regulatory pattern of "disruption and repair" (Beebe and Lachmann 1994), the "good-enough" caregiver who induces a stress response in her infant through a misattunement reinvokes in a timely fashion her psychobiologically attuned regulation of the infant's negative affect state *that she has triggered*. The reattuning, comforting mother and infant thus dyadically negotiate a stressful state transition of affect, cognition, and behavior. This recovery mechanism underlies the phenomenon of "interactive repair" (Tronick 1989), in which participation of the caregiver is responsible for the reparation of dyadic misattunements. In this process the mother who induces interactive stress and negative emotion in the infant is instrumental to the transformation of negative into positive emotion. It is now thought that "the process of reexperiencing positive affect following negative experience may teach a child that negativity can be endured and conquered" (Malatesta-Magai 1991, p. 218). Infant resilience is now being characterized as the capacity of *the child and the parent* to transition from positive to negative and back to positive affect (Demos 1991). It is important to note that resilience in the face of stress is an ultimate indicator of attachment capacity (Greenspan 1981).

These regulatory transactions underlie the formation of an attachment bond between the infant and primary caregiver. Attachment biology is thus the cement that provides what Freud (1916–1917) called the "adhesiveness" of early object relationships. Indeed, regulatory processes are now thought to be the precursors of psychological attachment and its associated emotions (Hofer 1994), and *psychobiological attunement is now understood to be the essential mechanism that mediates attachment bond formation* (Field 1985). In essence the baby becomes attached to the modulating caregiver who expands opportunities for positive affect and minimizes negative affect. In other words, *the affective state underlies and motivates attachment, and the central adaptive function of attachment dynamics is to interactively generate and maintain optimal levels of positive states and vitality affects.*

The Neurobiology and Psychobiology of Attachment Bond Formation

According to Bowlby (1969), vision is central to the establishment of a primary attachment to the mother, and imprinting is the learning mechanism that underlies attachment bond formation. Furthermore, attachment is more than overt behavior; it is internal, "being built into the nervous system, in the course and as a result of the infant's experience of his transactions with the mother" (Ainsworth 1967, p. 429). Emde (1988) asserts that the infant is biologically prepared to engage in visual stimulation in order to stimulate its brain. This brings us to another level of analysis—the neurobiological level. In this transfer of affect between mother and infant how are developing systems of the organizing brain influenced by these interactions with the social environment? Or, as Stechler and Halton (1987) put this question, what do we know of "the processes whereby the primary object relations become internalized and transformed into psychic structure" (p. 823)?

The work of Trevarthen (1993) on maternal-infant proto-conversations bears directly on this problem (see Figure 1–6). Coordinated with eye-to-eye messages are auditory vocalizations (tone of voice, "Motherese") as a channel of communication, and tactile and body gestures. A traffic of visual and prosodic auditory signals induces instant emotional effects,

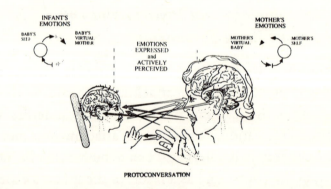

Figure 1–6: Channels of face-to-face communication in protoconversation. Protoconversation is mediated by eye-to-eye orientations, vocalizations, hand gestures, and movements of the arms and head, all acting in coordination to express interpersonal awareness and emotions (from Trevarthen 1993).

namely excitement and pleasure that build within the dyad. But Trevarthen also focuses on internal structure-function events. He points out that the engaged infant is interacting with the mother while she is in a state of *primary maternal preoccupation* (Winnicott 1956), and the resultant dyadic resonance ultimately permits the *intercoordination of positive affective brain states.* His work underscores the fundamental principle that the baby's brain is not only affected by these transactions, *its growth literally requires brain–brain interaction and occurs in the context of a positive affective relationship between mother and infant.* This interactive mechanism requires older brains to engage with mental states of awareness, emotion, and interest in younger brains, and involves a coordination between the motivations of the infant and the subjective feelings of adults.

Next question: What parts of the growing brain are affected by these events? I suggest that what is happening here is that the infant's right hemisphere, which is dominant for the child's processing of visual and prosodic emotional information and for the infant's recognition of maternal facial affective expressions, is psychobiologically attuned to the output of the mother's right hemisphere, which is involved in the expression and processing of emotional information and in spontaneous and nonverbal communication. The right cortex, which matures before the left, is known to be specifically impacted by early social experiences, to be activated in intense states of elation, and to contribute to the development of reciprocal interactions within the mother–infant regulatory system. The child is using the output of the mother's right cortex as a template for the imprinting, the hard wiring of circuits in his own right cortex, that will come to mediate his expanding affective capacities. It has been said that in early infancy the mother is the child's "auxiliary cortex" (Diamond et al. 1963). In these transactions she is "downloading programs" from her brain into the infant's brain. There is now solid evidence that the parenting environment influences the developing patterns of neural connections that underlie infant behavior (Dawson 1994).

Interactive transactions that regulate positive affect, in addition to producing neurobiological, structural consequences, are also generating important events at the psychobiological level. In describing the mother–infant experience of mutuality, Winnicott (1986) proposes that "the main thing is a communication between the baby and mother in terms of the anatomy and physiology of live bodies" (p. 258). This physiological linkage is an essential element of Kohut's (1971) postulate that the crucial maintenance of the infant's internal homeostatic balance is directly related to the infant's continuous dyadic reciprocal interactions with selfobjects. Indeed, self psychology is built upon a cardinal developmental principle: parents with mature psychological organizations serve as "selfobjects" that perform critical regulatory functions for the infant, who possesses an immature, incomplete psychological organization. This developmental psychoanalytic model is now being confirmed by Hofer's (1990, 1994) psychobiological research that demonstrates that in dyadic, symbiotic states the infant's open, immature, developing internal homeostatic systems are interactively regulated by the caregiver's more mature and differentiated nervous system. Selfobjects are thus external psychobiological regulators (Taylor 1987) that facilitate the regulation of affective experience (Palombo 1992), and they act at nonverbal levels beneath conscious awareness to co-create states of maximal cohesion and vitalization (Wolf 1988).

We now can understand the mechanism of Kohutian "mirroring." The human face is a unique stimulus for the display of biologically significant information. Current psychobiological studies of attachment show that in mutual gaze the mother's face is triggering high levels of endogenous opiates in the child's growing brain (Hoffman 1987, Panksepp et al. 1985). These endorphins, produced in the anterior pituitary, are biochemically responsible for the pleasurable qualities of social interaction and attachment as they act directly on dopamine neurons in the subcortical reward centers of the infant's brain that are responsible for heightened arousal (Schore 1994). Stimuli that induce arousal exert a potent influence on developmental processes (Rauschecker and Marler 1987). By promoting a symbiotic entrainment between the mother's mature and the infant's immature endocrine and nervous systems, hormonal responses are triggered that stimulate the child into a similar state of heightened CNS (Central Nervous System) arousal and sympathetic nervous system activity and resultant excitement and positive emotion. These findings support Basch's (1976) assertion that "the language of mother and infant consists of signals produced by the autonomic, involuntary nervous system in both parties" (p. 766).

In the latter half of the first year, object seeking, which Modell (1980) defines as the sharing and communicating of affects, specifically revolves around the mother's face, and it is her expressive face that is searched for and recognized (Wright 1991). High intensity mirroring exchanges thus create a merger experience that acts as a crucible for the forging of the affective ties of the attachment bond. This interactive mechanism creates what Mahler, Pine, and Bergman (1975) call "optimal mutual cueing, a perfect fit of the dual unity" (p. 204). Hofer (1990) states that "in postnatal life, the neural substrates for simple affective states are likely to be present and . . . the experiences for the building of specific pleasurable states are likewise built into the symbiotic nature of the earliest mother–infant interaction" (p. 62). *The concept of symbiosis is now solidly grounded in developmental research, and it should be returned into psychoanalysis.*

In his latest work Trevarthen (1993) refers to "The Self Born in Intersubjectivity: The Psychology of an Infant Communicating." He describes the emergence at 9–10 months of what he calls "secondary intersubjectivity." Indeed, in the last quarter of the first year the child's attachment experiences enable her now to share an intersubjective affect state with the caregiver (Lichtenberg 1989). Beebe and Lachmann (1994) conclude that facial mirroring allows for "entering into the other's changing feeling state" (p. 136). What's more, they assert that these experiences are stored in what they term *presymbolic representations*. These interactive representations appear at the end of the first year, and in them the infant represents the expectation of being matched by and being able to match the partner, as well as participating in the state of the other. This is also the identical time period when internal representations of working models of attachment are first encoded.

Attachment functions, which mature near the end of the first year of life, involve highly visual mechanisms and generate positive affect. These psychobiological experiences of attunement, misattunement, and re-attunement are imprinted into the early developing brain. In this manner "affects are at the crossroads of biology and history" (Modell 1984, p. 184). Stable attachment bonds that transmit high levels of positive affect are vitally important for the infant's continuing neurobiological development (Trad 1986). Mary Main (1991), perhaps the most influential current attachment researcher, now concludes that "the formation of an attachment to a specified individual signals a quantitative change in infant behavioral (*and no doubt also brain*) organization" (p. 214, italics added). Do we now know what parts of the brain begin a critical period of structural growth at 10 to 12 months and are involved in attachment functions and in the regulation of affect?

The Maturation of the Orbitofrontal Cortex during Mahler's Practicing Period (10–12 to 16–18 months)

In my book (Schore 1994) I offer evidence to show that dyadic communications that generate intense positive affect represent a growth-promoting environment for the prefrontal cortex, an area that is known to undergo a major maturational change at 10 to 12 months (Diamond and Doar 1989). It is now established that *frontal lobe functioning plays an essential role in the development of infant self-regulatory behavior* (Dawson et al. 1992). There is also evidence to show that, in particular, *orbital prefrontal areas* (see Figure 1–7) *are critically and directly involved in attachment functions* (Steklis and Kling 1985). This cortical area plays an essential role in the processing of social signals and in the pleasurable qualities of social interaction. Attachment experiences, face-to-face transactions between caregiver and infant, directly influence the imprinting, the circuit wiring of this system.

The orbital frontal cortex (so called because of its relation to the orbit of the eye) (Figure 1–8) is "hidden" in the ventral and medial surfaces of the prefrontal lobe (Figures 1–9 and 1–10) and acts as a convergence zone where cortex and subcor-

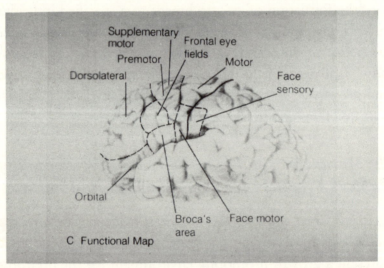

Figure 1–7: Approximate boundaries of functional zones of the human cerebral cortex, showing the dorsolateral and orbital prefrontal areas (from Kolb and Whishaw 1990).

tex meet (Figures 1–10 and 1–11). It sits at the hierarchical apex of the limbic system, the brain system responsible for the

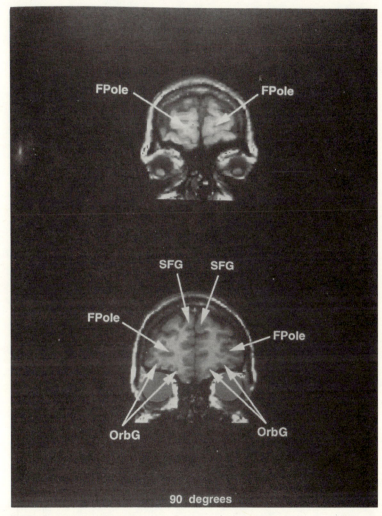

Figure 1–8: Computerized reconstruction of magnetic resonance (MR) imaging of a coronal section of the human brain. Notice orbital gyri (from H. Damasio 1995).

rewarding-excitatory and aversive-inhibitory aspects of emotion (Figure 1–12). This limbic cortex also acts as a major

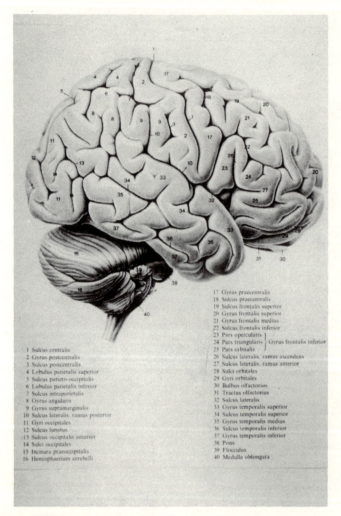

1 Sulcus centralis
2 Gyrus postcentralis
3 Sulcus postcentralis
4 Lobulus parietalis superior
5 Sulcus parieto-occipitalis
6 Lobulus parietalis inferior
7 Sulcus intraparietalis
8 Gyrus angularis
9 Gyrus supramarginalis
10 Sulcus lateralis, ramus posterior
11 Gyri occipitales
12 Sulcus lunatus
13 Sulcus occipitalis anterior
14 Sulci occipitales
15 Incisura praeoccipitalis
16 Hemisphaerium cerebelli

17 Gyrus praecentralis
18 Sulcus praecentralis
19 Sulcus frontalis superior
20 Gyrus frontalis superior
21 Gyrus frontalis medius
22 Sulcus frontalis inferior
23 Pars opercularis
24 Pars triangularis } Gyrus frontalis inferior
25 Pars orbitalis
26 Sulcus lateralis, ramus ascendens
27 Sulcus lateralis, ramus anterior
28 Sulci orbitales
29 Gyri orbitales
30 Bulbus olfactorius
31 Tractus olfactorius
32 Sulcus lateralis
33 Gyrus temporalis superior
34 Sulcus temporalis superior
35 Gyrus temporalis medius
36 Sulcus temporalis inferior
37 Gyrus temporalis inferior
38 Pons
39 Flocculus
40 Medulla oblongata

Figure 1–9: Lateral view of the human right hemisphere. Note the position of the orbital sulci (28) and gyri (29) in the frontal undersurface (from Nieuwenhuys et al. 1981).

control center over the sympathetic and parasympathetic branches of the ANS (Autonomic Nervous System), and thereby regulates drive and drive-restraint. Most significantly,

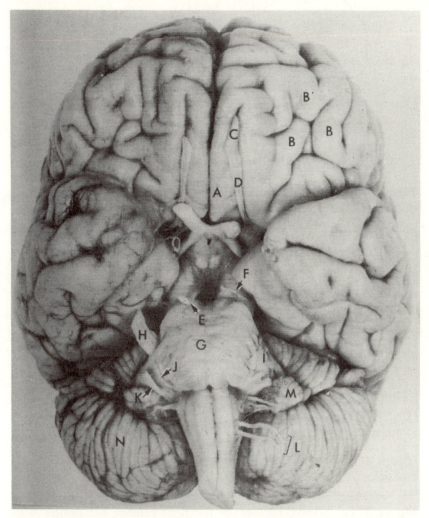

Figure 1–10: Photograph of the base of the human brain showing orbital gyri and sulci at sites labeled B (from Watson 1977).

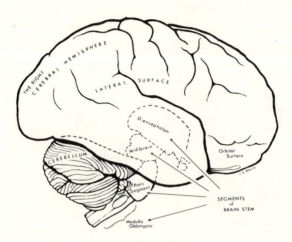

Figure 1–11: Relationships of brain stem structures to the orbital surface of the right hemisphere (from Smith 1981).

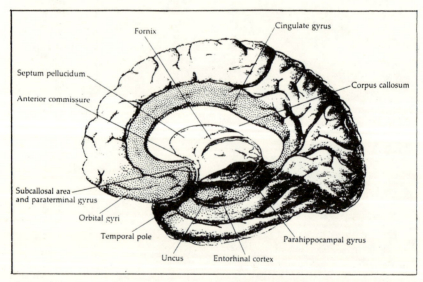

Figure 1–12: Midsagittal view of the right cerebral hemisphere with brain stem removed. The limbic association cortex is indicated by the dotted region. Note the orbital gyri (from Martin 1989).

in the cortex, *the orbitofrontal region is uniquely involved in social and emotional behaviors and in the homeostatic regulation of body and motivational states* (Schore 1994).

Due to its unique connections, at the orbitofrontal level cortically processed information concerning the *external* environment (such as visual and auditory stimuli emanating from the emotional face of the *object*) is integrated with sub-cortically processed information regarding the *internal* visceral environment (such as concurrent changes in the emotional or bodily *self* state). Neuroanatomists state that the function of this structure is involved with the internal state of the organism and is "closely tied to the synthesis of object-emotion relationships in a behavioral context" (Pandya and Yeterian 1990, p. 89). Orbitofrontal areas subserve memory (Stuss et al. 1982) and cognitive-emotional interactions (Barbas 1995), and are specialized to participate in the encoding of high-level, psychological representations of other individuals (Brothers and Ring 1992). This system thus contains the operational capacity to generate an internalized object relation, that is a self representation, an object representation, and a linking affect state (Kernberg 1976), or a Representation of Interactions that have been Generalized (RIGS) (Stern 1985).

The orbital prefrontal region is especially expanded in the right cortex (Falk et al. 1990), the hemisphere that is dominant for selectively attending to facial expressions. Because the early maturing and "primitive" right cortical hemisphere, more than the left, contains extensive reciprocal connections with limbic and subcortical regions, it is dominant for the

processing, expression, and regulation of emotional information (Joseph 1988). This hemisphere mediates pleasure and
pain (an essential concern of Freud's ideas about affect) and
the intrinsically more biologically primitive emotions that
serve fundamental motivational and social communication
functions. These primary emotions are nonverbal affects that
are spontaneously expressed on the face (Buck 1993). They
appear early in development, are expressed in universally rec-
ognizable configurations of facial movements, are correlated
with differentiable autonomic activity, and arise quickly and
"automatically." Autonomic nervous system control occurs
quite rapidly. It begins within one second and reaches full
development in 5 to 30 seconds. Automatic emotional processes are thus involuntary, effortless, and operate outside
conscious awareness.

This prefrontal region comes to act in the capacity of an
executive control function for the entire right cortex, the
hemisphere that modulates affect, nonverbal communication,
and unconscious processes. Early object relational experiences are not only registered in the deep unconscious, they
influence the development of the psychic systems that process
unconscious information for the rest of the lifespan. In this
manner, *the child's first relationship, the one with the mother,*
acts as a template for the imprinting of circuits in the child's
emotion-processing right brain, thereby permanently shaping
the individual's adaptive or maladaptive capacities to enter
into all later emotional relationships. Most intriguingly, the
activity of this non-dominant hemisphere, and not the later-
maturing dominant verbal-linguistic left, is instrumental to

the capacity of empathic cognition and the perception of the emotional states of other human beings (Voeller 1986). The right brain plays a superior role in the control of vital functions supporting survival and enabling the organism to cope actively and passively with stress and external challenge. Indeed, *the right brain is thought to contain the essential elements of the self system* (Mesulam and Geschwind 1978).

The right hemisphere contains an affective-configurational representational system, one that encodes self-and-object images unique from the lexical-semantic mode of the left (Watt 1990). According to Hofer (1984) *internal representations of external human interpersonal relationships serve an important intrapsychic role as "biological regulators" that control physiological processes.* These internal representations that contain information about state transitions (Freyd 1987) enable the child to self-regulate functions that previously required the caregiver's external regulation. There is now agreement that the encoding of strategies of affect regulation is a primary function of internal working models of attachment (Kobak and Sceery 1988) and that security of attachment fundamentally relates to a physiological coding that homeostatic disruptions will be set right (Pipp and Harmon 1987). Wilson and colleagues (1990) conclude that the experience of being with a self-regulating other is incorporated into an interactive representation. Of particular importance is the emergent capacity to access the complex symbolic representations that mediate evocative memory, since this allows for self-comforting during and subsequent to interactive stress (Fraiberg 1969).

Regulated and unregulated affective experiences with care-givers are thus imprinted and stored in early-forming proce-dural memory in the orbital prefrontal system and its cortical and subcortical connections as interactive representations. Current studies indicate that the development of parental representations and self representations occur in synchrony (Bornstein 1993), that internal representations develop epi-genetically through successive developmental stages (Blatt et al. 1990), and that developmental gradations of representa-tional capacity have important implications for affective development (Trad 1986). It is important to note that the infant's memory representation includes not only details of the learning cues of events in the external environment, but also of reactions in his internal state to changes in the exter-nal environment.

The orbitofrontal system, which Goleman (1995) calls "the thinking part of the emotional brain," plays a major role in the internal state of the organism (Mega and Cummings 1994), the temporal organization of behavior (Fuster 1985), and in the appraisal (Pribram 1987) and adjustment or cor-rection of emotional responses (Rolls 1986), that is, affect regulation. In fact it is one of the few brain regions that is "privy to signals about virtually any activity taking place in our beings' mind or body at any given time" (Damasio 1994, p. 181). It acts as a recovery mechanism that efficiently mon-itors and autoregulates the duration, frequency, and intensity of not only positive but negative affect states. This allows for a self-comforting capacity that can modulate distressing psy-chobiological states and reestablish positively toned states.

The essential activity of this psychic system is thus the adaptive switching of internal bodily states in response to changes in the external environment that are appraised to be personally meaningful. This emergent function, in turn, enables the individual to recover from disruptions of state and to integrate a sense of self across transitions of state, thereby allowing for a continuity of experience in various environmental contexts. These capacities are critical to the emergence, at 18 months, of a self system that is both stable *and* adaptable, a working definition of a dynamic system (M.D. Lewis 1995).

Infant observers report the emergence, at 18 months, of a "reflective self" that can take into account one's own and others' mental states, an achievement that is an essential step in emotional development (Fonagy et al. 1991). In the course of the second year the infant acquires the capacity to generate a "theory of mind," in which an individual imputes mental states to self and to others and predicts behavior on the basis of such states (Bretherton et al. 1981). The orbital cortex matures in the middle of the second year, the end of Mahler's practicing period, a time when the average child has a vocabulary of fifteen words. The core of the self is thus nonverbal and unconscious, and it lies in patterns of affect regulation.

Implications for Psychoanalytic Metapsychology

The nature and dynamics of this regulatory system that critically participates in the adaptive functions of mediating between the external environment and the internal milieu and in balancing external reality with internal desires is rele-

vant to the identification of psychoanalytic "psychic structure." In 1895 Freud hypothesized in his *"Project for a
Scientific Psychology"* that excitation from sources within and
outside of the individual might be regulated by processes
essentially within the individual (Schore in press). Over fifty
years ago, Hartmann (1939) proposed that development, in
essence, represents a differentiation in which primitive regulatory factors are increasingly replaced or supplemented by
more effective and adaptive regulatory factors. Twenty years
later he, along with Loewenstein (Hartmann and Loewenstein 1962), and then Schafer (1968) theorized that the
transformation of external regulations into internal ones
essentially defined the internalization process. In his last
work, Kohut (1984) concluded that early self–selfobject relationships allow for the maternally influenced creation of
structure involved in drive-regulating, integrative, and adaptive functions previously performed by the mother. Mahler
and colleagues (1975) described a psychic structural system
involved in the self-regulation of affect and therefore autonomous emotional functioning appears in the middle of the second year. And in recent work Settlage and colleagues (1988)
conceptualized development as a progression of stages in
which emergent adaptive self-regulatory structures and functions enable qualitatively new interactions between the individual and his environment.

Furthermore, this drive-modulating system is identical to
a controlling structure, described by Rapaport (1960) more
than thirty-five years ago, that maintains constancy by delay-
ing press for discharge of aroused drives. Holzman and Aron-

son (1992) have recently written that Freud "might have had some interest in contemporary neuropsychological studies of the frontal lobes in providing the organic infrastructure for channeling drives" (p. 72). In his current book, *Descartes' Error*, Damasio (1994), a neurological researcher, argues that emotions are "a powerful manifestation of drives and instincts," and emphasizes their motivational role: "In general, drives and instincts operate either by generating a particular behavior directly or by inducing physiological states that lead individuals to behavior in a particular way . . ." (p. 115). Recent psychobiological and neurobiological studies of emotion thus strongly indicate that *the concept of drive, a phenomenon at the frontier between the psychic and somatic, must be reintroduced as a central construct of psychoanalytic theory.*

The essential role of the orbitofrontal cortex in emotional-cognitive processes is now being explored by brain imaging techniques that allow us to image function as well as anatomy, to literally visualize images of mind and transient subjective states. For example, a positron emission tomography (PET) study demonstrates that when normal subjects silently fantasize dysphoric, affect-laden images of object loss, such as imagining the death of a loved one, increased blood flow and activation is recorded in the orbital prefrontal areas specifically (Pardo et al. 1993). In other words, we can now operationalize an on-line, real-time representation of an internal object relation. Interestingly, the PET scans of females reveal orbitofrontal activity in both hemispheres, while males show only unilateral activation, and more of the females than males experienced tearfulness during the imagery. An even more recent PET study also shows that women display significantly greater activity in this affect-regulating structure than men, especially in the right hemisphere (Andreasen et al. 1994). These data indicate gender differences in the wiring of the limbic system, and may relate to differences in empathic styles or capacities of processing nonverbal affect between the sexes. This may bear on the question of why nature has psychobiologically equipped women to be primary objects, and may be relevant to the "maternalization of psychoanalysis."

In another current functional neuroimaging study of introspective and self-reflective capacities, when subjects are asked to relax and listen to words that specifically describe what goes on in the mind (mental state terms such as wish, hope, imagine, desire, dream, and fantasy), a specifically increased activation of the right orbitofrontal cortex occurs (Baron-Cohen et al. 1994). And in a recently published PET study, Andreasen and colleagues (1995) report that during focused episodic memory, the recalling and relating of a personal experience to another, an increase of blood flow occurs in the orbitofrontal areas. Right frontal activity specifically occurs when the brain is actively retrieving this personal event from the past. Even more intriguingly, this same inferior frontal region is activated when the subject is told to allow the mind to "rest." In this condition of uncensored and silently unexpressed private thoughts, the individual's mental activity consists of loosely linked and freely wandering past recollections and future plans. The authors conclude that this orbitofrontal activity reflects "free association" that taps into primary process!

With regard to yet another aspect of primary process activity, Solms (1995), whose work on the organization of dreaming is at the interface of neurology and psychoanalysis, is now presenting neurological data that indicates that the control mechanism of dreaming is critically mediated by anterior limbic orbitofrontal structures. He concludes, "These regions are essential for affect regulation, impulse control, and reality testing; they act as a form of 'censorship'" (pp. 60–61). Normal activity in this brain system during sleep allows for the processing of information by symbolic representational mechanisms during dreaming, while failures in regulatory functioning caused by overwhelming experiences cause a breakdown in dreaming, disturbed sleep, and nightmares.

Implications for Psychoanalytic Conceptions of Psychopathology

There is now compelling evidence, from a number of separate disciplines that *all early forming psychopathology constitutes disorders of attachment and manifests itself as failures of self and/or interactional regulation* (Grotstein 1986). According to Grotstein (1990) the fundamental pathology of these patients, who frequently manifest some form of neurobiological impairment, traces back to a failure of good enough attachment and bonding. As a consequence of an inability to gain access to maternal modulation of their affective states, they experience a lifelong "inability to self-regulate, to receive, encode, and process the data of emotional experience they are subjected to" (p. 157). Furthermore, it now appears that the "lack of differentiation between self and other results because of the semipermeability of physiological regulation between infant and mother" (Pipp 1993, p. 194).

Indeed, borderline personalities exhibit an inability to self-regulate and lack the capacity to form "stable" self and object representations (Grotstein 1987). Narcissistic personalities, although developmentally more advanced, also exhibit insecure attachments (Pistole 1995) and manifest an impairment of self-esteem regulation and a disturbance in the representation of self in relationship to others (Auerbach 1990). Such personalities never developmentally attain a psychic organization that can generate complex symbolic representations that contain information about transitioning out of stress-induced negative states. Instead they frequently access pathological internal representations that encode a dysregulated self-in-interaction-with-a-misattuning-other.

I propose that the functional indicators of this adaptive limitation are specifically manifest in recovery deficits of internal reparative mechanisms. This psychopathology is manifest in a limited capacity to modulate the intensity and duration of affects, especially biologically primitive affects like shame, rage, excitement, elation, disgust, and panic-terror, and it may take the form of either under- or over-regulation disturbances. Such deficits in coping with negative affects are most obvious under challenging conditions that call for behavioral flexibility and adaptive responses to socioemotional stress. This conceptualization fits well with recent models that emphasize that loss of ability to regulate the intensity of feelings is the most far-reaching effect of early trauma and neglect (van der Kolk and Fisler 1994) and that this dysfunction is manifest in more intense and longer-lasting emotional responses (Oatley and Jenkins 1992). I suggest that these *functional vulnerabilities* reflect *structural weaknesses* and defects in the organization of the orbitofrontal cortex, the neurobiological regulatory structure that is centrally involved in the adjustment or correction of emotional responses.

For some time, psychoanalysis has speculated that developmental disorders reflect a defect in an internal psychic structure, but it has been unable to identify this structural system. I believe that the orbital prefrontolimbic system is this psychic structure, and further, that every type of early-forming disorder involves, to some extent, altered orbital prefrontal limbic function (Schore 1996). Anatomical studies highlight the unique developmental plasticity of the prefrontal limbic cortex, and this property has been suggested to mediate its "preferential vulnerability" in psychiatric disorders (Barbas 1995). Indeed, very recent research (mostly brain-imaging studies over the last year) shows evidence for impaired orbitofrontal activity in such diverse severe psychopathologies as autism (Baron-Cohen 1995), schizophrenia (Seidman et al. 1995), mania (Starkstein et al. 1988), unipolar depression (Mayberg et al. 1994), phobic states (Rauch et al. 1995), posttraumatic stress disorder (Semple et al. 1992), drug addiction (Volkow et al. 1991), alcoholism (Adams et al. 1995), and borderline (Goyer et al. 1994), and psychopathic (Lapierre et al. 1995) personality disorders.

Implications for Psychoanalytic Clinical Theory

A theoretical perspective centered in developmental psychoanalysis, psychobiology, and neurobiology emphasizes the salience of early affective phenomena, and the emergent paradigm created by this interdisciplinary integration also has significant implications for clinical models of psychoanalysis, whose focus of primary study is now being described as "human emotional development and functioning" (Langs and Badalamenti 1992). Indeed, "affect theory is increasingly recognized as the most likely candidate to bridge the gap between clinical theory and general theory in psychoanalysis" (Spezzano 1993, p. 39). Krystal (1988) underscores the fundamental principle that the development and maturation of affects represents the key event in infancy. In a recent volume Knapp (1992) argues that "optimal regulation is a goal of maturation, including therapeutic maturation. Dysregulation is a therapeutic target" (p. 247). The psychotherapy of "developmental arrests" (Stolorow and Lachmann 1980) is currently conceptualized as being directed toward the mobilization of fundamental modes of development (Emde 1990) and the completion of interrupted developmental processes (Gedo 1979).

In early preverbal development, the infant constructs internal working models of the attachment relationship with his caregivers, and these representations, permanently imprinted into maturing brain circuitries, determine the individual's characteristic approach to affect modulation for the rest of the lifespan. The restoring into consciousness and reassessment of these internalized working models, suggested by Bowlby (1988) to be the essential task of psychoanalytic psychotherapy is identical to Kernberg's (1988) assertion that unconscious, nonverbally communicated "units constituted by a self-representation, an object-representation, and an affect state linking them are the essential units of psychic structure relevant for psychoanalytic exploration" (p. 482). These interactive representations are stored in the right hemisphere that contains an affective-configurational representational system and is dominant for the processing of emotional information.

The direct relevance of developmental studies to the psychotherapeutic process derives from the commonality of interactive emotion transacting mechanisms in the caregiver–infant relationship and in the therapist–patient relationship. The essence of development is contained in the concept of "reciprocal mutual influences" (Schore 1996). Early preverbal maternal–infant emotional communications that occur before the maturation of the left hemisphere and the onset of verbal-linguistic capacities represent contingently responsive affective transactions between the right hemispheres of the members of the dyad. Yet, the "non-verbal, prerational stream of expression that binds the infant to its parent continues throughout life to be a primary medium of intuitively felt affective-relational communication between persons" (Orlinsky and Howard 1986, p. 343). A young infant functions in a fundamentally unconscious way, and unconscious processes in an older child or adult can be traced back to the primitive functioning of the infant (Fischer and Pipp 1984). Unconscious processes are, of course, most clearly revealed and expressed in transference–countertransference interactions that are characterized, according to Stolorow and colleagues (1987), by "mutual reciprocal influence." Freud's (1913) statement that "everyone possesses in his own unconscious an instrument with which he can interpret the utterances of the unconscious of other people" (p. 320) and Racker's (1968) discovery that "every transference situation provokes a countertransference situation" (p. 137) clearly imply a transactional model of the psychotherapeutic relationship.

In a recently published work, Eisenstein and colleagues (1994) describe the importance, frequency, and variety of nonverbal communications that take place between therapist and patient. These communications are expressed in tone of voice, facial expression, and body posture, outside the realm of awareness of both, yet transference and countertransference reactions occur in response to these cues. Indeed the authors conclude that this nonverbal exchange represents a major factor in the therapeutic process. Demos (1992), from the perspective of infant research, cautions us that psychoanalysis "overvalues verbal and symbolic modes of representation and undervalues nonverbal and presymbolic modes. It places language in a privileged position as the only reliable source of information about the inner experience of another" (p. 208). In a recent paper on the critical role of nonverbal components in the clinical psychoanalytic process, Jacobs (1994) states that mastering the art of understanding nonverbal communication is "like learning to become an accomplished baby watcher" (p. 748). Krystal (1988) notes that the "infantile nonverbal affect system" continues to operate throughout life. I conclude that *a deeper understanding of the interactive affect transacting mechanisms of the nonverbal, unconscious transference–countertransference relationship represents the frontier of clinical psychoanalysis.*

There is now a growing consensus that despite the existence of a number of distinct *theoretical* perspectives in psychoanalysis, the *clinical* concepts of transference (Wallerstein 1990) and countertransference (Gabbard 1995) represent a common ground. The deeper elucidation of the mechanisms that underlie these core phenomena therefore becomes an important goal. In my book I present multidisciplinary evidence to support the proposal that *nonverbal transference–countertransference interactions that take place at preconscious-unconscious levels represent right hemisphere to right hemisphere communications of fast-acting, automatic, regulated, and unregulated emotional states between patient and therapist.* The orbitofrontal cortex, intimately involved with internal, bodily, and motivational states, plays a preeminent role in this interactive mechanism. In fact, studies now show that it functionally mediates the capacity of empathizing (Mega and Cummings 1994) and inferring the states of others (Baron-Cohen 1995), and of reflecting on one's own internal emotional states, as well as others' (Povinelli and Preuss 1995). Furthermore, this cortical area is essentially involved in the control of the allocation of attention to possible contents of consciousness (Goldenberg et al. 1989). This system is expanded in the right cerebral cortex that is responsible for the manifestations of emotional states (Ross 1984) and unconscious processes (Galin 1974, Watt 1990).

Implications for a Psychoanalytic Model of Psychotherapy

A model of treatment that integrates developmental and clinical perspectives generates heuristic hypotheses about the underlying dynamic mechanisms that are involved in the psychotherapeutic experience, especially those involved in the primitive emotional disorders that characterize the early-forming right hemispheric impairments of preoedipal developmental psychopathologies. Of special importance to the psychotherapy process are early-forming representations of a dysregulated-self-in-interaction-with-a-misattuning-other, since these become unconscious templates of emotional relationships that mediate psychopathology. Such representations are imprinted predominantly with painful primitive affect that the developmentally impaired personality cannot intrapersonally or interpersonally regulate. As a result of this limitation, certain forms of external and internal affect-inducing input are selectively and defensively excluded from conscious processing.

A body of clinical and experimental evidence now indicates all forms of psychopathology have concomitant symptoms of emotional dysregulation, and that defense mechanisms are, in essence, forms of emotional regulation strategies for avoiding, minimizing, or converting affects that are too difficult to tolerate (Cole et al. 1994). However, it is

just these strategies of affect regulation and pathogenic schemas of dysregulation that must be recognized and addressed
in the transference–countertransference matrix. Such "latent"
"sequestered" schemas are, according to Slap and Slap-Shelton (1994) egocentric, analogical, and visual. They are stored
in the visuospatial right hemisphere, which contains an analogical representational system (Tucker 1992) and a nonverbal processing mode that are inaccessible to the language
centers in the left (Joseph 1982). From this realm that stores
split-off parts of the self also come transference projections
that are directed outward into the therapist.

*The pathology of early-forming developmental disorders is
most clearly revealed under conditions of interpersonal stress.*
Though early painful experiences are buried in deep layers of
the unconscious, during stress their effects are felt on the surface, especially at the interface where the self interacts with
other selves, selves who are potential sources of dysregulation. A prime example of this occurs during the sudden rupture of the therapeutic alliance that accompanies the negative
therapeutic reaction, a regressive worsening of the patient's
condition following what seems to be adequate therapeutic
management. "When one speaks hopefully to them or
expresses satisfaction with the progress of treatment, they
show signs of discontent and their condition invariably
becomes worse" (Freud 1923, p. 39). I suggest that in actuality this represents the therapist's misattunement to the
patient's current state.

How might we understand the rapid disorganizing events
of this clinical phenomenon? It is now thought that critical

cues generated by the therapist, which are absorbed and metabolized by the patient, generate the transference (Gill 1982), an "activation of existing units of internalized object relations" (Kernberg 1980). These cues resemble the parents' original toxic behavior at heightened affective moments of misattunement, and they are processed by the patient's right hemisphere, which is preferentially activated under stress conditions (Tucker et al. 1977). Of particular importance are visual and auditory cues that were perceived during early self-disorganizing episodes of shame-humiliation, a common element of borderline and narcissistic histories (Schore 1991).

Facial indicators of transference processes (Krause and Lutolf 1988) are quickly appraised from the therapist's face in movements occurring primarily in the regions around the eyes and from prosodic expressions from the mouth (Fridlund 1991) by the patient's right cortical mechanisms involved in the perception of nonverbal expressions embedded in facial and prosodic stimuli (Blonder et al. 1991). Such input generates "a series of analogical comparisons between distortions by the therapist ('misalliance') and the empathic failures and distortions of parents" (Watt 1986, p. 61). This instantly activates right brain imprinted pathological internal object relations and "hot cognitions" (Greenberg and Saffran 1984), which program the patient's "hot theory of mind" (Brothers 1995), which constructs others' evaluative attitudes and meaningful intentions. These early interactive representations encode expectations of imminent dysregulation. As a result the patient's brain suddenly shifts dominance from a mode of left hemispheric linear processing to right hemispheric nonlinear processing.

Indeed, the right hemisphere is involved in the memorial storage of emotional faces (Suberi and McKeever 1977) and is activated during the recall of autobiographical (Cimino et al. 1991) and early childhood memories (Horowitz 1983, Joseph 1992). The current interactive stress, similar in form to a very early misattuned, dysregulating transaction, instantly ruptures the attachment bond between patient and therapist. This sudden shattering of the therapeutic alliance thus represents a reconstruction of what Lichtenberg (1989) calls a "model scene," and it induces the entrance into consciousness of a chaotic state associated with early traumatic experiences that is stored in implicit (Siegel 1995) procedural memory and usually protected by "infantile amnesia." But now, due to state-dependent recall (Bower 1981) the patient is propelled into a bodily state that psychobiologically designates a "dreaded state of mind" (Horowitz 1987), thereby triggering splitting, the instant evaporation of the positive and sudden intensification of the negative transference. This malignant transference reaction, manifest in rapid emotional activation and instability, reflects hyperarousal- or hypoarousal-associated alterations of limbic regions (McKenna 1994). As a result of the subsequent rapid escalation of intense negative affect the self disorganizes, either explosively or implosively. Neurobiological studies show that the emergence of strong affect during psychotherapy is accompanied by increased right hemispheric activation (Hoffman and Goldstein 1981).

It is important to remember that *this affective state is transmitted within the dyad*. The therapist's resonance with this right brain state, in turn, triggers "somatic countertransference" (Lewis 1992) and these "somatic markers" (Damasio 1994) may be physiological responses that receive (or block) the patient's distress-inducing projective identifications. Sander (1992) refers to a "mutuality of influence . . . a thinking that is oriented as much around the way the patient's signals influence therapist state as around therapist on patient state" (p. 583). This dialectical mechanism is especially prominent during stressful ruptures of the working alliance that occur during "enactments," defined as those "events occurring within the dyad that both parties experience as being the consequence of behavior in the other" (McLaughlin 1991, p. 611). The rapid-onset, dynamic events of the "negative therapeutic reaction" are thus an overt manifestation of the interaction of the patient's covert, deep, unconscious transference patterns with the clinician's covert, deep, unconscious countertransference patterns.

The working-through process at this point is a dyadic venture of interactive repair (Tronick 1989) and depends very much upon *the therapist's ability to recognize and regulate the negative affect within himself*. Ellman (1991) points out that although the handling of negative affect and the negative transference is the most difficult part of treatment, the therapist's tolerance and containment of negative states is an important contributor to the creation of analytic trust. The clinician's participation in the disruption and repair process (Beebe and Lachmann 1994) is dependent upon and limited by his or her capacity to tolerate and cope with the patient's negative state which *he or she has (unconsciously) triggered.* This coping capacity is reflected in an ability, *under stress,* to self-regulate (contain) the projected negative affect, and thereby act as an interactive regulator of the shared negative state. In doing so, he resonates with the patient's internal state of arousal dysregulation, modulates it, *communicates it back* prosodically in a more regulated form, and then verbally labels his state experiences.

The essential step in this process is the therapist's ability, initially at a nonverbal level, to detect, recognize, monitor, and self-regulate the countertransferential stressful alterations in his bodily state that are evoked by the patient's transferential communication. In doing so, the therapist must engage in a "reparative withdrawal," a self-regulating maneuver that allows continued access to a state in which a symbolizing process can take place, thereby enabling him to create a parallel affective and imagistic scenario that resonates with the patient's (Freedman and Lavender in press). According to these authors the presence or absence of the therapist's recognition of his countertransferential bodily signals and his capacity to autoregulate the disruption in state caused by the patient literally determine whether or not the countertransference is destructive or constructive, symbolizing or desymbolizing.

Thus, the active involvement of *both members of the dyad* in the process of disruption and repair is absolutely necessary for the patient's learning that a previously self-disorganizing state can be regulated (rather than further dysregulated) by an external object. The patient can now, in the presence of a reparative object, transition out of a previously avoided stressful state into one in which he can associate the nonverbal affect state with verbal processing. Researchers are currently showing that "the ability to express oneself in words during states of high emotional arousal is an important achievement in self-regulation" (Dawson 1994, p. 358). Wolf (1991) holds that as a result of a successful repair process the mutual reciprocal empathic bond between patient and therapist becomes stronger and less vulnerable to repeated disruptions. And Gedo (1995) now writes that working through, "the difficult transitional process whereby reliance on former modes of behavioral regulation is gradually superseded by more effective adaptive measures," is accomplished by "the mastery of affective intensities" (p. 344).

Unconscious affect and its regulation thus becomes a primary goal of the psychotherapy of preoedipal dynamics, especially a focus on the recognition and identification of affects that were never developmentally interactively regulated or internally represented. Therapeutic interventions are directed toward the elevation of emotions from a primitive, presymbolic, sensorimotor level of experience to a mature, symbolic, representational level, a functional advance that is mediated by an increased flexibility of the patient's emotional control structures. In long-term work, with the internalization of the therapist's regulatory functions, the patient becomes capable of accessing a self-reflective position that can appraise the significance and meanings of a variety of emotional states. As Bach (1985) points out, this developmental achievement is expressed in the emergence of a higher-level integrative capacity that allows "free access to affective memories of alternate states, a kind of superordinate reflective awareness that permits multiple perspectives on the self" (p. 179).

These functional advances reflect alterations in internal structures. In the current neuropsychiatric literature Mender (1994) writes that "psychoanalytic recall, through a reacquaintance with the most primitive and undifferentiated sources of human potential, can rejuvenate our range of neurobiological options" (p. 169). I suggest that the mobilization of fundamental modes of development that occurs in psychotherapy reflects the organization of structural alterations in limbic circuitries that neurobiologically mediate the emergence of adaptive capacities. In psychoanalytic writings Basch (1988) asserts that psychotherapy can facilitate the alteration and reworking of the patterns in the patient's nervous system that govern how he/she processes socioemotional information. In fact it is now thought that specifically cortical and sensorilimbic connections are reworked in long-term dynamic psychotherapy (McKenna 1994).

In concordance with this model, Spezzano (1993) proposes that

> the analytic relationship heals by drawing into itself those methods of processing and regulating affect relied on by the patient for psychological survival and then transforming them. The mechanism of these transformations is the regulation of affect in a better way within the analysis than it was previously managed by the patient and the subsequent modification of what, in the classical language of

structural change, might be called the patient's unconscious affect-regulating structures. [pp. 215–216]

Watt (1986) is even more specific—he speculates that the
connections of the right frontolimbic cortex, a neurobiological structure involved in the regulation of primitive affects,
are specifically reorganized by the psychoanalytic experience.
Most intriguingly, these hypotheses about the nature of the
internal structural system that is altered in psychotherapy has
recently been corroborated in a PET imaging study that
demonstrates that patients show significant changes in metabolic activity in the right orbitofrontal cortex and its subcortical connections as a result of successful psychological
treatment (Schwartz et al. 1996).

These results are supported by a large body of studies in
the neurosciences that indicates that although the effects of
environmental experiences develop more rapidly and extensively in the developing than the adult brain, the capacity for
experience-dependent plastic changes in the nervous system
remains throughout the lifespan, and indeed, experience is
necessary for the full growth of brain and behavioral potential
(Rosenzweig 1996). In fact, there is now evidence that *the
prefrontal limbic cortex, more than any other part of the cerebral cortex, retains the plastic capacities of early development.*
The orbitofrontal cortex, even in adulthood, continues to
express anatomical and biochemical features observed in
ontogeny, and this accounts for its great plasticity and
involvement in learning, memory, and cognitive-emotional
interactions (Barbas 1995). Such findings suggest that this

particular system, with its capacity for utilizing and directing the psychobiological expression of learning encoded within the limbic system (Rossi 1993), is a critical site of the psychic structural changes that are a product of a long-term, growth-facilitating psychotherapeutic relationship.

References

Adams, K. M., Gilman, S., Koeppe, R., et al. (1995). Correlation of neuropsychological function with cerebral metabolic rate in subdivisions of the frontal lobes of older alcoholic patients measured with [^{18}F] fluorodeoxyglucose and positron emission tomography. *Neuropsychology* 9:275–280.

Ainsworth, M. D. S. (1967). *Infancy in Uganda: Infant Care and the Growth of Love.* Baltimore: Johns Hopkins University Press.

Andreasen, P. J., O'Leary, D. S., Cizadlo, T., et al. (1995). Remembering the past: two facets of episodic memory explored with positron emission tomography. *American Journal of Psychiatry* 152:1576–1585.

Andreasen, P. J., Zametkin, A. J., Guo, A. C., et al. (1994). Gender-related differences in regional cerebral glucose metabolism in normal volunteers. *Psychiatry Research* 51:175–183.

Auerbach, J. S. (1990). Narcissism: reflections on others' images of an elusive concept. *Psychoanalytic Psychology* 7:545–564.

Bach, S. (1985). *Narcissistic States and the Therapeutic Process.* New York: Jason Aronson.

Barbas, H. (1995). Anatomic basis of cognitive-emotional interactions in the primate prefrontal cortex. *Neuroscience and Biobehavioral Reviews* 19:499–510.

Baron-Cohen, S., (1995). *Mindblindness: An Essay on Autism and Theory of Mind.* Cambridge, MA: MIT Press.

Baron-Cohen, S., Ring, H., Moriarty, J., et al. (1994). Recognition of mental state terms: clinical findings in children with autism and a functional neuroimaging study of normal adults. *British Journal of Psychiatry* 165:640–649.

Basch, M. F. (1976). The concept of affect: a re-examination. *Journal of the American Psychoanalytic Association* 24:759–777.

———— (1988). *Understanding Psychotherapy.* New York: Basic Books.

Beebe, B., and Lachmann, F. M. (1988). Mother–infant mutual influence and precursors of psychic structure. In *Progress in Self Psychology, vol. 3*, ed. A. Goldberg, pp. 3–25. Hillsdale, NJ: Analytic Press.

———— (1994). Representations and internalization in infancy: three principles of salience. *Psychoanalytic Psychology* 11:127–165.

Blank, H. R. (1975). Reflection on the special senses in relation to the development of affect with special emphasis on blindness. *Journal of the American Psychoanalytic Association* 23:32–50.

Blatt, S. J., Quinlan, D. M., and Chevron, E. (1990). Empirical investigations of a psychoanalytic theory of depression. In *Empirical Studies of Psychoanalytic Theories, vol. 3*, ed. J. Masling, pp. 89–147. Hillsdale, NJ: Analytic Press.

Blonder, L. X., Bowers, D., and Heilman, K. M. (1991). The role of the right hemisphere in emotional communication. *Brain* 114:1115–1127.

Bornstein, R. F. (1993). Parental representations and psychopathology: a critical review of the empirical literature. In *Psychoanalytic Perspectives on Psychopathology*, ed. J. M. Masling and R. F. Bornstein, pp. 1–41. Washington DC: American Psychological Association.

Bower, G. H. (1981). Mood and memory. *American Psychologist* 36:129–148.

Bowlby, J. (1969). *Attachment and Loss, vol 1: Attachment.* New York: Basic Books.

———— (1988). Attachment, communication, and the therapeutic process. In *A Secure Base: Clinical Applications of Attachment Theory.* London: Routledge.

Bretherton, I., McNew, S., and Beeghly, M. (1981). Early person knowledge in gestural and verbal communication: When do infants acquire a "theory of mind"? In *Infant Social Cognition*, ed. M. Lamb and L. Sherrod, pp. 335–373. Hillsdale, NJ: Lawrence Erlbaum.

Brothers, L. (1995). Neurophysiology of the perception of intentions by primates. In *The Cognitive Neurosciences*, ed. M. S. Gazzaniga, pp. 1107–1115. Cambridge, MA: MIT Press.

Brothers, L., and Ring, B. (1992). A neuroethological framework for the representations of minds. *Journal of Cognitive Neuroscience* 4:107–118.

Buck, R. (1993). Spontaneous communication and the foundation of the interpersonal self. In *The Perceived Self: Ecological and Interpersonal Sources of Self-knowledge*, ed. U. Neisser, pp. 216–236. New York: Cambridge University Press.

Cimino, C. R., Verfaellie, M., Bowers, D., and Heilman, K. M. (1991). Autobiographical memory: influence of right hemisphere damage on emotionality and specificity. *Brain and Cognition* 15:106–118.

Cole, P. M., Michel, M. K., and O'Donnell Teti, L. (1994). The development of emotion regulation and dysregulation: a clinical perspective. *Monographs of the Society for Research in Child Development* 59:73–100.

Cooper, A. M. (1987). Changes in psychoanalytic ideas: transference interpretation. *Journal of the American Psychoanalytic Association* 35:77–98.

Damasio, A. R. (1994). *Descartes' Error.* New York: Grosset/Putnam.

Damasio, H. (1995). *Human Brain Anatomy in Computerized Images.* New York: Oxford University Press.

Dawson, G. (1994). Development of emotional expression and emotion regulation in infancy. In *Human Behavior and the Developing Brain*, ed. G. Dawson and K. W. Fischer, pp. 346–379. New York: Guilford.

Dawson, G., Panagiotides, H., Klinger, L. G., and Hill, D. (1992). The role of frontal lobe functioning in the development of infant self-regulatory behavior. *Brain and Cognition* 20:152–175.

Demos, V. (1991). Resiliency in infancy. In *The Child in Our Times: Studies in the Development of Resiliency*, ed. T. F. Dugan and R. Coles. New York: Brunner/Mazel.

——— (1992). The early organization of the psyche. In *Interface of Psychoanalysis and Psychology*, ed. J. W. Barron, M. N. Eagle, and D. L. Wolitsky, pp. 200–232. Washington DC: American Psychological Association.

Diamond, A., and Doar, B. (1989). The performance of human infants on a measure of frontal cortex function, the delayed response task. *Developmental Psychobiology* 22:271–294.

Diamond, M. C., Krech, D., and Rosenzweig, M. R. (1963). The effects of an enriched environment on the histology of the rat cerebral cortex. *Journal of Comparative Neurology* 123:111–120.

Eisenberg, L. (1995). The social construction of the human brain. *American Journal of Psychiatry* 152:1563–1575.

Eisenstein, S., Levy, N. A., and Marmor, J. (1994). *The Dyadic Transaction: An Investigation into the Nature of the Psychotherapeutic Process.* New Brunswick, NJ: Transaction.

Ellman, S. J. (1991). *Freud's Technique Papers: A Contemporary Perspective.* Northvale, NJ: Jason Aronson.

Emde, R. N. (1988). Development terminable and interminable. I. Innate and motivational factors from infancy. *International Journal of Psycho-Analysis* 69:23–42.

—— (1990). Mobilizing fundamental modes of development: empathic availability and therapeutic action. *Journal of the American Psychoanalytic Association* 38:881–913.

Falk, D., Hildebolt, C., Cheverud, J., et al. (1990). Cortical asymmetries in frontal lobes of rhesus monkeys *(Macaca mulatta). Brain Research* 512:40–45.

Feldman, R., and Greenbaum, C. W. (1997). Affect regulation and synchrony in mother–infant play as precursors to the development of symbolic competence. *Infant Mental Health Journal* 18:4–23.

Feldman, R., Greenbaum, C. W., Yirmiya, N., and Mayes, L. C. (1996). Relations between cyclicity and regulation in mother–infant interaction at 3 and 9 months and cognition at two years. *Journal of Applied Developmental Psychology* 17:347–365.

Field, T. (1985). Attachment as psychobiological attunement: being on the same wavelength. In *The Psychobiology of Attachment and Separation,* ed. M. Reite and T. Field, pp. 415–454. Orlando, FL: Academic Press.

Field, T., and Fogel, A. (1982). *Emotion and Early Interaction.* Hillsdale, NJ: Lawrence Erlbaum.

Fischer, K. W., and Pipp, S. L. (1984). Development of the structures of unconscious thought. In *The Unconscious Reconsid-*

ered, ed. K. S. Bowers and D. Meichenbaum, pp. 88–148. New York: Wiley.

Fonagy, P., Steele, M., Steele, H., et al. (1991). The capacity for understanding mental states: the reflective self in parent and child and its significance for security of attachment. *Infant Mental Health Journal* 12:201–218.

Fraiberg, S. (1969). Libidinal object constancy and mental representation. *Psychoanalytic Study of the Child* 24:9–47. New York: International Universities Press.

Fraiberg, S., and Freedman, D. A. (1964). Studies in the ego development of the congenitally blind. *Psychoanalytic Study of the Child* 19:113–169. New York: International Universities Press.

Freedman, N., and Lavender, J. (in press). On receiving the patient's transference: the symbolizing and desymbolizing countertransference. *Journal of the American Psychoanalytic Association*.

Freud, S. (1895). Project for a scientific psychology. *Standard Edition* 1:281–397.

———— (1913). The claims of psycho-analysis to scientific interest. *Standard Edition* 13:161–190.

———— (1916–1917). Introductory lectures on psycho-analysis. *Standard Edition* 15/16.

———— (1923). The ego and the id. *Standard Edition* 19.

Freyd, J. J. (1987). Dynamic mental representations. *Psychological Reviews* 94:427–438.

Fridlund, A. (1991). Evolution and facial action in reflex, social motive, and paralanguage. *Biological Psychology* 32:3–100.

Fuster, J. M. (1985). The prefrontal cortex and temporal integration. In *Cerebral Cortex. vol. 4. Association and Auditory Cortices*, ed. A. Peters and E. G. Jones, pp. 151–171. New York: Plenum.

Gabbard, G. O. (1995). Countertransference: the emerging common ground. *International Journal of Psycho-Analysis* 76:475–485.

Galin, D. (1974). Implications for psychiatry of left and right cerebral specialization: a neuropsychological context for unconscious processes. *Archives of General Psychiatry* 31:572–583.

Gedo, J. (1979). *Beyond Interpretation.* New York: International Universities Press.

———— (1995). Working through as metaphor and as a modality of treatment. *Journal of the American Psychoanalytic Association* 43:339–356.

Gill, M. M. (1982). *Analysis of Transference.* New York: International Universities Press.

Goldenberg, G., Podreka, I., Uhl, F., et al. (1989). Cerebral correlates of imagining colours, faces and a map—I. SPECT of regional cerebral blood flow. *Neuropsychologia* 27:1315–1328.

Goleman, D. (1995). *Emotional Intelligence.* New York: Bantam.

Goyer, P. F., Konicki, P. E., and Schulz, S. C. (1994). Brain imaging in personality disorders. In *Biological and Neurobehavioral Studies of Borderline Personality Disorder*, ed. K. R. Silk, pp. 109–125. Washington, DC: American Psychiatric Press.

Greenberg, L. S., and Safran, J. D. (1984). Hot cognition: emotion coming in from the cold. A reply to Rachman and Mahooney. *American Psychologist* 44:19.

Greenspan, S. I. (1981). *Psychopathology and Adaptation in Infancy and Early Childhood.* New York: International Universities Press.

Grotstein, J. S. (1986). The psychology of powerlessness: disorders of self-regulation and interactional regulation as a newer paradigm for psychopathology. *Psychoanalytic Inquiry* 6:93–118.

———— (1987). The borderline as a disorder of self-regulation. In *The Borderline Patient: Emerging Concepts in Diagnosis*, ed. J. S. Grotstein, J. Lang, and M. Solomon, pp. 347–383. London: Analytic Press.

———— (1990). Invariants in primitive emotional disorders. In *Master Clinicians on Treating the Regressed Patient*, ed. L. B. Boyer and P. L. Giovacchini, pp. 139–163. Northvale, NJ: Jason Aronson.

Hartmann, H. (1939). *Ego Psychology and the Problem of Adaptation*. New York: International Universities Press.

Hartmann, H., and Loewenstein, R. M. (1962). Notes on the superego. *Psychoanalytic Study of the Child* 17:42–81. New York: International Universities Press.

Hess, E. H. (1975). *The Tell-Tale Eye*. New York: Van Nostrand Reinhold.

Himwich, W. A. (1975). Forging a link between basic and clinical research: developing brain. *Biological Psychiatry* 10:125–139.

Hobson, R. P. (1993). Through feeling and site through self and symbol. In *The Perceived Self: Ecological and Interpersonal Sources of Self-knowledge*, ed. U. Neisser, pp. 254–279. New York: Cambridge University Press.

Hofer, M. A. (1984). Relationships as regulators: a psychobiologic perspective on bereavement. *Psychosomatic Medicine* 46:183–197.

———— (1990). Early symbiotic processes: hard evidence from a soft place. In *Pleasure Beyond the Pleasure Principle*, ed. R. A. Glick and S. Bone, pp. 55–78. New Haven: Yale University Press.

———— (1994). Hidden regulators in attachment, separation, and loss. *Monographs of the Society for Research in Child Development* 59:192–207.

Hoffman, E., and Goldstein, L. (1981). Hemispheric quantitative EEG changes following emotional reactions in neurotic patients. *Acta Psychiatrica Scandinavica* 63:153–164.

Hoffman, H. S. (1987). Imprinting and the critical period for social attachments: some laboratory investigations. In *Sensitive Periods in Development: Interdisciplinary Studies,* ed. M. H. Bornstein, pp. 99–121. Hillsdale, NJ: Lawrence Erlbaum.

Holzman, P., and Aronson, G. (1992). Psychoanalysis and its neighboring sciences: paradigms and opportunities. *Journal of the American Psychoanalytic Association* 40:63–88.

Horowitz, M. J. (1983). *Image Formation and Psychotherapy.* New York: Jason Aronson.

—— (1987). *States of Mind: Configurational Analysis of Individual Psychology.* New York: Plenum Medical Book Company.

Jacobs, T. J. (1994). Nonverbal communications: some reflections on their role in the psychoanalytic process and psychoanalytic education. *Journal of the American Psychoanalytic Association* 42:741–762.

Joseph, R. (1982). The neuropsychology of development: hemispheric laterality, limbic language, and the origin of thought. *Journal of Clinical Psychology* 38: 4-33.

—— (1988). The right cerebral hemisphere: emotion, music, visual-spatial skills, body-image, dreams, and awareness. *Journal of Clinical Psychology* 44:630–673.

—— (1992). *The Right Brain and the Unconscious: Discovering the Stranger Within.* New York: Plenum.

Keeler, W. R. (1958). Autistic patients and defective communication in blind children with retrolental fibroplasia. In *Psychopa-*

thology of Communication, ed. P. H. Hoch and J. Zubin. New York: Grune & Stratton.

Kernberg, O. (1976). *Object Relations and Clinical Psychoanalysis*. New York: Jason Aronson.

—— (1980). *Internal World and External Reality*. New York: Jason Aronson.

—— (1988). Object relations theory in clinical practice. *Psychoanalytic Quarterly* 57:481–504.

Kobak, R. R., and Sceery, A. (1988). Attachment in late adolescence: working models, affect regulation, and representations of self and others. *Child Development* 59:135–146.

Kohut, H. (1971). *The Analysis of the Self.* New York: International Universities Press.

—— (1984). *How Does Analysis Cure?* Chicago: University of Chicago Press.

Knapp, P. H. (1992). Emotion and the psychoanalytic encounter. In *Affect: Psychoanalytic Perspectives,* ed. T. Shapiro and R. N. Emde, pp. 239–264. Madison, CT: International Universities Press.

Kolb, B., and Whishaw, I. Q. (1990). *Fundamental of Human Neuropsychology,* 3rd ed. New York: W. H. Freeman.

Krause, R., and Lutolf, P. (1988). Facial indicators of transference processes within psychoanalytic treatment. In *Psychoanalytic Process Research Strategies,* ed. H. Dahl and H. Kachele. New York: Springer-Verlag.

Krystal, H. (1988). *Integration and Self-healing: Affect-Trauma-Alexithymia*. Hillsdale, NJ: Analytic Press.

Langs, R., and Badalamenti, A. (1992). The three modes of the science of psychoanalysis. *American Journal of Psychotherapy* 46:163–182.

Lapierre, D., Braun, C. M. J., and Hodgins, S. (1995). Ventral frontal deficits in psychopathy: neuropsychological test findings. *Neuropsychologia* 33:139–151.

Lewis, M. D. (1995). Cognition-emotion feedback and the self-organization of developmental paths. *Human Development* 38:71–102.

Lewis, P. (1992). The creative arts in transference–countertransference relationships. *The Arts in Psychotherapy* 19:317–323.

Lichtenberg, J. D. (1989). *Psychoanalysis and Motivation*. Hillsdale, NJ: Analytic Press.

Mahler, M., Pine, F., and Bergman, A. (1975). *The Psychological Birth of the Human Infant*. New York: Basic Books.

Main, M. (1991). Discourse, prediction, and recent studies in attachment: implications for psychoanalysis. *Journal of the American Psychoanalytic Association* 41:209–244, suppl.

Malatesta-Magai, C. (1991). Emotional socialization: its role in personality and developmental psychopathology. In *Internalizing and Externalizing Expressions of Dysfunction: Rochester Symposium on Developmental Psychopatholgy, vol. 2* , ed. D. Cicchetti and S. L. Toth, pp. 203–224. Hillsdale, NJ: Lawrence Erlbaum.

Martin, J. H. (1989). *Neuroanatomy: Text and Atlas*. New York: Elsevier.

Mayberg, H. S., Lewis, P. J., Regenold, W., and Wagner, H. N., Jr. (1994). Paralimbic hypoperfusion in unipolar depression. *Journal of Nuclear Medicine* 35:929–934.

McKenna, C. (1994). Malignant transference: a neurobiologic model. *Journal of the American Academy of Psychoanalysis* 22:111–127.

McLaughlin, J. T. (1991). Clinical and theoretical aspects of enactment. *Journal of the American Psychoanalytic Association* 39:595–614.

Mega, M. S., and Cummings, J. L. (1994). Frontal-subcortical circuits and neuropsychiatric disorders. *Journal of Neuropsychiatric and Clinical Neuroscience* 6:358–370.

Mender, D. (1994). *The Myth of Neuropsychiatry: A Look at Paradoxes, Physics, and the Human Brain.* New York: Plenum.

Mesulam, M.-M., and Geschwind, N. (1978). On the possible role of the neocortex and its limbic connections in the process of attention in schizophrenia: clinical cases of inattention in man and experimental anatomy in monkey. *Journal of Psychiatric Research* 14:249–259.

Modell, A. H. (1980). Affects and their non-communication. *International Journal of Psycho-Analysis* 61:259–267.

——— (1984). *Psychoanalysis in a New Context.* New York: International Universities Press.

Nagera, H., and Colonna, A. (1965). Aspects of the contribution of sight to ego and drive development—a comparison of the development of some blind and sighted children. *Psychoanalytic Study of the Child* 20:267–287. New York: International Universities Press.

Nieuwenhuys, R., Voogd, J., and van Huijzen, C. (1981). *The Human Central Nervous System: A Synopsis and Atlas,* 2nd ed. rev. New York: Springer-Verlag.

Oatley, K., and Jenkins, J. M. (1992). Human emotions: function and dysfunction. *Annual Review of Psychology* 43:55–85.

Orlinsky, D. E., and Howard, K. I. (1986). Process and outcome in psychotherapy. In *Handbook of Psychotherapy and Behavior Change, 3rd ed.,* ed. S. L. Garfield and A. E. Bergin, pp. 311–381. New York: Wiley.

Palombo, J. (1992). Narratives, self-cohesion, and the patient's search for meaning. *Clinical Social Work Journal* 20:249–270.

Pandya, D. N., and Yeterian, E. H. (1990). Prefrontal cortex in relation to other cortical areas in rhesus monkey: architecture and connections. *Progress in Brain Research* 85:63–94.

Panksepp, J., Siviy, S. M., and Normansell, L. A. (1985). Brain opioids and social emotions. In *The Psychobiology of Attachment and Separation*, ed. M. Reite and T. Field, pp. 3–49. Orlando, FL: Academic Press.

Pardo, J. V., Pardo, P. J., and Raichle, M. E. (1993). Neural correlates of self-induced dysphoria. *American Journal of Psychiatry* 150:713–718.

Pipp, S. (1993). Infant's knowledge of self, other, and relationship. In *The Perceived Self*, ed. U. Neisser, pp. 185–204. New York: Cambridge University Press.

Pipp, S., and Harmon, R. J. (1987). Attachment as regulation: a commentary. *Child Development* 58:648–652.

Pistole, M. C. (1995). Adult attachment style and narcissistic vulnerability. *Psychoanalytic Psychology* 12:115–126.

Povinelli, D., and Preuss, T. M. (1995). Theory of mind: evolutionary history of a cognitive specialization. *Trends in Neuroscience* 18:418–424.

Preisler, G. M. (1995). The development of communication in blind and in deaf infants—similarities and differences. *Child: Care, Health and Development* 21:79–110.

Pribram, K. H. (1987). The subdivisions of the frontal cortex revisited. In *The Frontal Lobes Revisited,* ed. E. Perecman, pp. 11–39. Hillsdale, NJ: Lawrence Erlbaum.

Racker, H. (1968). *Transference and Countertransference.* New York: International Universities Press.

Rapaport, D. (1960). The structure of psychoanalytic theory. [*Psy-*

chological Issues, Monograph 6]. New York: International Universities Press.

Rauch, S. C., Savage, C. R., Alpert, N. M., et al. (1995). A positron emission tomographic study of simple phobic symptom provocation. *Archives of General Psychiatry* 52:20–28.

Rauschecker, J. P., and Marler, P. (1987). Cortical plasticity and imprinting: behavioral and physiological contrasts and parallels. In *Imprinting and Cortical Plasticity: Comparative Aspects of Sensitive Periods*, pp. 349–366. New York: Wiley.

Rolls, E. T. (1986). Neural systems involved in emotion in primates. In *Emotion: Theory, Research, and Practice, vol. 3.*, ed. R. Plutchik and H. Kellerman, pp. 125–143. Orlando, FL: Academic Press.

Rosenzweig, M. R. (1996). Aspects of the search for neural mechanisms of memory. *Annual Review of Psychology* 47:1–32.

Ross, E. D. (1984). Right hemisphere's role in language, affective behavior and emotion. *Trends in Neuroscience* 7:342–346.

Rossi, E. L. (1993). *The Psychobiology of Mind–Body Healing*. New York: Norton.

Sander, L. W. (1992). Letter to the editor. *International Journal of Psycho-Analysis* 73:582–584.

Schafer, R. (1968). *Aspects of Internalization*. New York: International Universities Press.

Schore, A. N. (1991). Early superego development: the emergence of shame and narcissistic affect regulation in the practicing period. *Psychoanalysis and Contemporary Thought* 14:187–250.

——— (1994). *Affect Regulation and the Origin of the Self: The Neurobiology of Emotional Development*. Hillsdale, NJ: Lawrence Erlbaum.

——— (1996). The experience-dependent maturation of a regulatory system in the orbital prefrontal cortex and the origin of developmental psychopathology. *Development and Psychopathology* 8:59–87.

——— (in press). One hundred years after Freud's *Project for a Scientific Psychology*—Is a rapprochement between psychoanalysis and neurobiology at hand? *Journal of the American Psychoanalytic Association.*

——— (work in progress). *Affect Regulation and the Repair of the Self.* New York: Guilford.

Schwartz, J. M., Stoessel, P. W., Baxter, L. R., Jr., et al. (1996). Systematic cerebral glucose metabolic rate changes after successful behavior modification treatment of obsessive-compulsive disorder. *Archives of General Psychiatry* 53:109–113.

Seidman, L. J., Oscar-Berman, M., Kalinowski, A. G., et al. (1995). Experimental and clinical neuropsychological measures of prefrontal dysfunction in schizophrenia. *Neuropsychology* 9:481–490.

Semple, W. E., Goyer, P., McCormick, R., et al. (1992). Increased orbital frontal cortex blood flow and hippocampal abnormality in PTSD: a pilot PET study. *Biological Psychiatry* 31:129A.

Settlage, C. F., Curtis, J., Lozoff, M., et al. (1988). Conceptualizing adult development. *Journal of the American Psychoanalytic Association* 36:347–369.

Siegel, D. J. (1995). Memory, trauma, and psychotherapy: a cognitive science view. *Journal of Psychotherapy Practice and Research* 4:93–122.

Slap, J. W., and Slap-Shelton, L. J. (1994). The schema model: a proposed replacement paradigm for psychoanalysis. *Psychoanalytic Review* 81:677–693.

Smith, C. G. (1981). *Serial Dissection of the Human Brain.* Baltimore and Munich: Urban and Scwarzenberg.

Solms, M. (1995). New findings on the neurological organization of dreaming: implications for psychoanalysis. *Psychoanalytic Quarterly* 64:43–67.

Spezzano, C. (1993). *Affect in Psychoanalysis: A Clinical Synthesis.* Hillsdale, NJ: Analytic Press.

Spitz, R. A. (1958). On the genesis of superego components. *Psychoanalytic Study of the Child* 13:375–404. New York: International Universities Press.

Starkstein, S. E., Boston, J. D., and Robinson, R. F. (1988). Mechanisms of mania after brain injury: 12 case reports and review of the literature. *Journal of Nervous and Mental Disease* 176:87–100.

Stechler, G., and Halton, A. (1987). The emergence of assertion and aggression during infancy: a psychoanalytic systems approach. *Journal of the American Psychoanalytic Association* 35:821–838.

Steklis, H. D., and Kling, A. (1985). Neurobiology of affiliative behavior in nonhuman primates. In *The Psychobiology of Attachment and Separation*, ed. M. Reite and T. Field, pp. 93–134. Orlando, FL: Academic Press.

Stern, D. N. (1977). *The First Relationship.* Cambridge, MA: Harvard University Press.

—— (1983). Early transmission of affect: some research issues. In *Frontiers of Infant Psychiatry*, ed. J. Call, E. Galenson, and R. Tyson, pp. 52–69. New York: Basic Books.

—— (1985). *The Interpersonal World of the Infant.* New York: Basic Books.

Stolorow, R. D., and Lachmann, F. M. (1980). *Psychoanalysis of Developmental Arrests*. New York: International Universities Press.

Stolorow, R. D., Brandchaft, B., and Atwood, G. (1987). *Psychoanalytic Treatment: An Intersubjective Approach*. Hillsdale, NJ: Analytic Press.

Stuss, D. T., Kaplan, E. F., Benson, D. F., et al. (1982). Evidence for the involvement of orbitofrontal cortex in memory functions: an interference effect. *Journal of Comparative Physiological Psychology* 96:913–925.

Suberi, M., and McKeever, W. F. (1977). Differential right hemispheric memory storage of emotional and non-emotional faces. *Neuropsychologia* 15:757–768.

Taylor, G. (1987). *Psychosomatic Medicine and Contemporary Psychoanalysis*. Madison, CT: International Universities Press.

Tomkins, S. (1962). *Affect/Imagery/Consciousness: Vol. 1, The Positive Affects*. New York: Springer.

Trad, P. V. (1986). *Infant Depression*. New York: Springer-Verlag.

Trevarthen, C. (1993). The self born in intersubjectivity: the psychology of an infant communicating. In *The Perceived Self: Ecological and Interpersonal Sources of Self-knowledge*, ed. U. Neisser, pp. 121–173. New York: Cambridge University Press.

Tronick, E. Z. (1989). Emotions and emotional communication in infants. *American Psychologist* 44:112–119.

Tucker, D. M. (1992). Developing emotions and cortical networks. In *Minnesota Symposium on Child Psychology. Vol. 24, Developmental Behavioral Neuroscience*, ed. M. R. Gunnar and C. A. Nelson, pp. 75–128. Hillsdale, NJ: Lawrence Erlbaum.

Tucker, D. M., Roth, R. S., Arneson, B. A., and Buckingman, V. (1977). Right hemisphere activation during stress. *Neuropsychologia* 15:697–700.

van der Kolk, B. A., and Fisler, R. E. (1994). Childhood abuse and neglect and loss of self-regulation. *Bulletin of the Menninger Clinic* 58:145–168.

Voeller, K. K. S. (1986). Right-hemisphere deficit syndrome in children. *American Journal of Psychiatry* 143:1004–1009.

Volkow, N. D., Fowler, J. S., Wolf, A. P., et al. (1991). Changes in brain glucose metabolism in cocaine dependence and withdrawal. *American Journal of Psychiatry* 148:621–626.

Wallerstein, R. (1990). Psychoanalysis: the common ground. *International Journal of Psycho-Analysis* 71:3–19.

Watson, C. (1977). *Basic Human Neuroanatomy, An Introductory Atlas,* 2nd ed. Boston: Little, Brown.

Watt, D. F. (1986). Transference: A right hemispheric event? An inquiry into the boundary between psychoanalytic metapsychology and neuropsychology. *Psychoanalysis and Contemporary Thought* 9:43–77.

—— (1990). Higher cortical functions and the ego: explorations of the boundary between behavioral neurology, neuropsychology, and psychoanalysis. *Psychoanalytic Psychology* 7:487–527.

Wilson, A., Passik, S. D., and Faude, J. P. (1990). Self-regulation and its failures. In *Empirical Studies of Psychoanalytic Theory, vol. 3,* ed. J. Masling, pp. 149–213. Hillsdale, NJ: Analytic Press.

Winnicott, D. W. (1956). Primary maternal preoccupation. In *Through Paediatrics to Psychoanalysis,* ed. J. D. Sutherland, pp. 300–305. The International Psycho-Analytical Library. London: Hogarth.

——— (1958). The capacity to be alone. *International Journal of Psycho-Analysis* 39:416–420.

——— (1986). *Home is Where We Start From.* New York: Norton.

Wolf, E. S. (1988). *Treating the Self: Elements of Clinical Self Psychology.* New York: Guilford.

——— (1991). Advances in self psychology: the evolution of psychoanalytic treatment. *Psychoanalytic Inquiry* 11:123–146.

Wright, K. (1991). *Vision and Separation: Between Mother and Baby.* Northvale NJ: Jason Aronson.

2

The Significance of the First Few Months of Life for Self-Regulation: A Reply to Schore

Steven J. Ellman and Catherine Monk

Dr. Schore's work marks the fulfillment of a dream held by a number of psychoanalysts: the tying together of biological—particularly neurophysiological—data with analytic theory and experience. His work demonstrates how psychoanalysis can remain true to its theoretical foundation and to the phenomenology of the analytic process while drawing on other disciplines to enrich it. Thus, a hermeneutic or a narrative perspective becomes simply one way to approach the analytic process—useful sometimes, at other times perhaps not—and is not the philosophically exclusive domain in which one can analyze the human condition, as some of its stronger proponents would have it.

Dr. Schore's remarkable integration begins with the early mother–infant relationship as it occurs on three interacting levels: internal, behavioral, and neurobiological. His synthesis assimilates a wider area than that, however, as he considers the ramifications of his findings for analytic theory and technique. Drawing extensively on two domains not often considered in relation to one another—cutting-edge mother–child and neurobiological research—Dr. Schore offers a model of the development of affect regulation grounded in attachment processes, mapped out on the neurological plane, and relevant to psychopathology and clinical practice. He also offers a model of how dyadic *process* becomes intra-individual *structure* that incorporates and satisfies behavioral, neurobiological, and psychoanalytic perspectives. His is a breathtaking project.

This commentary will begin by questioning Dr.

Schore's entry point in the mother–child story (seemingly after three months, with an emphasis on the interactions of an older infant) and his focus on visual communication over other modes. With these questions and the discussion of additional mother–child data, we will suggest that an earlier mother–child experience could be usefully inserted into Dr. Schore's model. Such an addition could correct some inaccuracies we find in his graphing of psychoanalytic theory onto the behavioral and neurobiological processes of the mother–child relationship as well as address his problematic interchanging of concepts such as *affect regulation* and *self-regulation* as well as *homeostasis, arousal* and *drive*. We also will argue against Schore's version of Freudian theory. To be sure, Freudian theory must be updated to accommodate his model of mind-brain relationships but we will attempt to show that the way it is updated falls within Freud's essential postulates. Finally, our conceptualization of early mother–child experience will be briefly related to clinical technique.

Dr. Schore's discussion of the mother–child dyad opens with his focus on visual communication. He cites neurobiological data showing the increased myelination of the visual area when the infant is 2 to 3 months old and refers to many mother–child gaze-gaze aversion interaction studies dating from the "second and third quarter of the first year" (when the infant is between 4 and 9 months old). When Dr. Schore next discusses "the psychobiological experiences of attunement, mis-attunement, and re-attunement [that] are imprinted into the early developing brain," he bases these matches and mismatches of communication on visual mechanisms which, he claims, mature near the end of the first year, when the infant is 10 to 12 months old. Finally, he maps the dyadic visual communication process, which he suggests is the primary medium for the development of attachment and affect regulation, onto the orbitofrontal cortex, which he says matures significantly when the infant is between 10 and 12 months old.

From the behavioral level to the neurobiological one, it seems that Dr. Schore is overlooking the infant's first few months of life and, consequently, the ramifications of these first months for the child's developing brain and regulatory capacities. Furthermore, Dr. Schore highlights visual communication between mother and baby to the near exclusion of other modes of contact such as touch, voice, smell, taste, and sense of bodily contact. While obviously central to mother–child communication and the child's emotional development (both on behavioral and neurobiological levels), it is as if Dr. Schore's decision to rely on visual communication as his primary theoretical building block led him to overlook the younger, less visually active baby, and the mother–child interaction of that earlier period. Dr. Schore's specificity regarding visual communication raises questions about the congenitally blind as well. According to his model, the congenitally blind must, of necessity, be severely psychologically impaired. Clearly we know that this is not the case in the gross form implied in Dr. Schore's paper; there are some congenitally blind people who have made secure attachments and have adequate means of self regulation. Other non-visual modes of interaction, which are also primary forms of communication, must serve as alternative pathways to the typical visual mode of later development in the extraordinary adaptation of some blind people.

We hypothesize, in contrast to Dr. Schore, that before an infant can rely on vision in its establishment of a primary attachment to the mother and development of emotional regulation, the baby must satisfactorily pass through an earlier stage where taste, touch, smell, sound, and *physiological equilibrium* are more central. We hope to use behavioral and neurobiological data to support our hypotheses as well as to show that such a formulation is more consistent with psychoanalytic theory than Dr. Schore's model.

Touch, according to Brazelton and Cramer (1990), is "the first important area of communication between a mother and her new infant" (p. 61). Touch is a "message system" between caregiver and infant that can lead to quietness or arousal. Whereas a slow patting motion is soothing, more rapid patting becomes an alerting stimulus and the threshold for arousal is "very specific" (p. 61). Infants naturally mold their body to an adult's and come equipped with the rooting reflex whereby a touch to the mouth causes them to turn in the direction of the stimulus—an ideal mechanism for finding a nipple (Brazelton 1992). Six-day-old babies can reliably distinguish the odor of their mother's breast pads from those of other lactating mothers (MacFarlane 1975). Not only can newborns detect subtle differences in taste, but they seem to be programmed with a burst–pause sucking pattern that they reserve for breast feeding. Researchers interpret this pattern as suggesting that babies expect more social interaction when being fed at the breast than with a bottle (Brazelton and Cramer 1990). Newborn babies prefer human to nonhuman sounds (Brazelton and Cramer 1990), and Condon and Sander (1975) have shown that immediately after birth newborns will synchronize their movements to the rhythm of a mother's voice. Finally, while a 3-week-old baby can recognize his or her mother's face, for about the first two months of life, babies are nearsighted and prefer to look at faces and objects ten to twelve inches in front of them, an ideal situation for keeping the caregiver physically close by (Brazelton and Cramer 1990).

This very brief synopsis suggests that infants come into the world primed to establish a specific attachment relationship, to keep that significant other very close by, and are able, through responsiveness in sucking patterns as well as other indications of pleasure and pain, to motivate the adult to attune to them and their needs. As an example of the adult's responsiveness, consider that a mother can differentiate her baby's cry from that of other newborns by the third day and that she can distinguish pain, hunger, and boredom cries by the end of the second week (Lester and Zeskind 1982). During this early part of life, the caregiver is attuning to the child in order to help him or her obtain, maintain, and regain physiological equilibrium. The mother and child, are, in a sense, communicating and mutually adapting to each other for the purpose of titrating the child's state of arousal and needs. When infants encounter too much stimulation, they will do what researchers call *habituation*. They will close down their nervous system, respond less and less to the stimuli, and appear to be asleep, although they will actually be in a different state that includes tightened flexed extremities and no eye blinks. It is as if these babies are actively maintaining control over their environment rather than relaxing into sleep. Habituation is an important means of self-protection and suggests active regulation of state control (Brazelton and Cramer 1990).

During this first part of life, particularly the first few weeks, an awake, alert state that lasts for twenty to thirty minutes is considered long. There is a predominance of REM sleep at this time, which has led to the idea that in this state brain growth and differentiation occur (Brazelton and Cramer 1990). In addition, one of us (SE), with a variety of colleagues, has conducted research postulating that the function of REM sleep is to provide endogenous stimulation, a hypothesis that sheds further light on the infant's early regulatory processes.

This sleep research charts the relationship between REM sleep and intracranial self-stimulation (a procedure in which an electrode is inserted into an animal's brain and the animal presses a lever to get the electrode to stimulate this brain site and produce an experience of pleasure). The research shows that if an animal is deprived of REM sleep its threshold for intracranial self-stimulation (ICSS) is greatly lowered and it will choose to obtain greater amounts of ICSS (Steiner and Ellman 1972). There is thus a reciprocal relationship: if an animal is deprived of REM, it desires more ICSS; if an animal is allowed ICSS, REM is abolished with no subsequent rebound (Ellman et al. 1991, Spielman et al. 1973). This work suggests that the onset of REM sleep activates pleasure sites. What importance do these studies have for human infants? Ellman (1992) and Jouvet (1967) have postulated that the function of REM sleep is the facilitation of responses that the infant will need for survival. This means that mechanisms that underlie food seeking, sexual and aggressive behavior, and, in the human (probably in all primates), various social behaviors that bond infant and mother, are experienced and, so to speak, rehearsed in REM sleep. Thus, in REM there are periodic smiling and various facial expressions (that will leave or rather be actively inhibited after six months) as well as sucking behavior and erection cycles. In the first month of life the infant is forming what might be termed passive and active bonding precursors. These are stimulated increasingly by external sources (primarily the mother) but at the beginning of life by endogenous stimulation inherent in REM sleep. REM sleep is central to providing the baby endogenous stimulation at a time when it is not overly receptive to large amounts of exogenous stimulation.

Infants come into the world equipped to capture the caregiver's attention and direct it to their physiological state. The caregiver watches over their rest/activity cycle, trying to make the transitions as smooth as possible. At times when infants are overwhelmed, they habituate to shut out the external stimulation on their own. And although they spend a majority of their time in REM sleep, during this state they experience stimulation and pleasure.

What this look at the first few months of life suggests is that important experiences in regulation, particularly physiological regulation, are taking place *prior to* the entry point of Dr. Schore's model. This period of early regulation, which is significantly shaped by the quality of early caregiving, must also have important ramifications for the infant's brain development and later self regulation processes. Our assertion is that Dr. Schore's model does not pay enough attention to the earliest mother–child experiences that specifically influence the child's sense of state management of external as well as internal stimuli.

It is here that the question of terms emerges. Dr. Schore seems to use *affect regulation, self-regulation, homeostatic regulation* and *affect, arousal,* and *drive* somewhat interchangeably. While there is not the space to consider these various terms—only to say that we would reserve the term *affect* to those feeling states that serve a signaling function both interpersonally and intrapsychically—they are distinct concepts. Self-regulation, which refers to arousal and homeostasis and can include drive experiences, may underlie or contribute to affect regulation, but it is questionable whether these different concepts should be collapsed into one. Furthermore, what are the consequences of raising affect to this central position in psychoanalytic theory and technique, above others?

In describing his model, Dr. Schore elegantly moves among behavioral data, neurological research, and psychoanalytic theory. However, we believe his overlooking of the first few months of life has resulted in some inaccurate uses of psychoanalytic theory as well as a misperception that Freud's model cannot encompass the new behavioral and neurological research. It is to these comments we now turn.

Jacobson (1964) theorized that during the earliest months of life the mother's concerns centered on the baby's physiological processes. She invented the term *organ language* to describe this period of development. Winnicott, perhaps more than any other theorist, elucidated this early mother–child relationship. He felt that there was "no such thing as an infant," only an infant and mother unit (1960). During the early period, which he termed *absolute dependence*, the infant is *held*, a word he used to denote the total environmental provision the baby receives. Taking seriously Freud's libido theory and his assertion that "we do not say of objects which serve the interests of self-preservation that we love them, we emphasize the fact that we *need* them" (Freud 1915, p. 138, italics added), Winnicott saw mothers as meeting infants' needs emanating from within and without. Through her holding, the mother offers the infant ego support so that an emerging self can process the self's experience of an instinct, lest it be disorganizing (Winnicott 1960). She protects the infant from physiological insults and impingements and, in so doing, promotes his or her development from a state of unintegration to one of integration. At this point in life the baby is concerned with pleasure–unpleasure sequences and the mother is not so much mirroring the baby as finding the baby's needs through anticipation and finding herself in the baby. What the baby must have, in Winnicott's (1960) words, is "*a live adaptation to [his or her] needs*" (p. 54, italics original).

Winnicott's and Jacobson's (as well as others') theoretical

ideas of early development are consistent with the behavioral and sleep research we briefly reviewed. In particular, Winnicott's ideas dovetail with a focus on the mother's role in the infant's regulatory development that takes place during the first few months of life. We were thus struck by Dr. Schore's use of the Winnicottian terms *going on being* and *primary maternal preoccupation*. While Winnicott was never precise in his time frame, he was clear that going on being and primary maternal preoccupation are very early experiences in the mother–child unit that take place when the child is unaware of the outside other managing his or her experience *and is younger than three months old* (Winnicott 1956, 1960). Again, we suggest that an underemphasis on the first few months of life has led to a missed opportunity to accurately integrate psychoanalytic developmental theory with behavioral and neuropsychological data of the first few months of life.

This brings us to Freudian theory in general. Dr. Schore suggests that "Freud's structural theory must not be abandoned but updated in terms of what we know about brain-mind relationships." As one of us has argued (Ellman 1995), Freud had four separate models of the mind. One of them, his object relations model, is the one from which Winnicott, Klein, and even Kohut borrow heavily and is more receptive to the type of integration Dr. Schore makes in his paper. In many of Freud's works during his object relations phase (1911–1919), he describes an early phase of autoerotism in which the pleasure ego is in existence. In this period, the infant is unaware of the regulatory care being given it. This first phase of Freud's theory is consistent with our description of the first

few months of life. The second phase in Freud's object relations model of the mind can be called narcissism, within which
he postulated a time when all good things (including aspects of
the mother) would be seen as inside the infant and all bad
(including aspects of the mother) would be represented as outside the infant's self. Freud (1915) referred to this new, narcissistic self organization as the Purified Pleasure Ego (PPE).
(Few have noted that this is the concept with which Kohut
begins his description of narcissistic development.) Freud, in
several long footnotes (as Winnicott has pointed out), realized
that he was again not including a crucial aspect of the baby's
experience, namely the baby's mother. However, Freud's concern was to develop a conceptual framework for how he saw
the infant developing and representing the world. Freud
almost always started from *inside* the infant in terms of his theoretical focus and did not include the mother at this time
because the baby was unaware of her as separate.

We concur with Dr. Schore that concepts such as mirroring, selfobject, and symbiosis thus have their place in describing developmental experiences from the infant's perspective
even when we, from the outside, can see two separate beings
mutually cueing each other. Freud's object relations model of
the mind, specifically his ideas of narcissism and the PPE, are
entirely compatible with Dr. Schore's thesis of the development of attachment and regulation and with his data. However, while Freud's face, "the visual symbol of classical
psychoanalysis," may, as a caricature of his theory, suggest "a
monad, a single unit," this is an enervation of his ideas. Even
in "The Ego and the Id" (1923), the work that epitomizes the

structural theory (another of his models of the mind), Freud
sees the character of the ego as "a precipitate of abandoned
object cathexes" and states that it "contains the history of
those object-choices" (p. 29)—hardly a theory of a monadical
existence but one that speaks to the object relation, the dyadic
process, that went into the development of one being.

As a last point, we will comment on Dr. Schore's recommendations for treatment that are derived from his ideas
about the development of affect regulation. Dr. Schore states
that there should be a moment-to-moment tracking of "hot
cognitions" that trigger nonlinear phenomenological discontinuities. He asserts, like others (Bach 1994, Ellman 1991,
Steingart 1995), that treatment should, it is hoped, progress
from a presymbolic sensorimotor level of experience to a
more mature representational level of symbolization. However, Dr. Schore's stress is on affect regulation, while we
think that the stress should be moved over slightly to affective capacity, that is, the capacity to experience, endure, and
regulate affect within a self structure that contains as few dis-
associated elements as possible.

In previous work, one of us (Ellman 1991, 1996) has
identified the first task of analytic treatment as helping
patients begin to create a new reality in the therapeutic situation. This occurs when patients begin to feel listened to and
understood in a way that is unlike what they have previously
known. A necessary condition for this sense of a new reality is
analytic trust (Ellman 1991), which involves the therapist's
entering the patient's world not as an impingement but rather
as a person attuned to the patient, particularly in terms of the

depth, intensity, and lability of emotional states. Empathically resonating with a patient's affective state and consistently reflecting and synthesizing his or her experiences are the building blocks of analytic trust as well as the experiential means to facilitate the patient's readiness for interpretations and tolerance of his or her own affective experiences. Before one can interpret in treatment, it is important to help patients see how from moment to moment they try to rid themselves of emotions that are particularly aversive or threaten to disrupt the stable experience of sense of self or object. This emphasis on affective capacity draws on an understanding of the *earliest* mother–child experience. While not incompatible with Dr. Schore's ideas, the psychobiological perspective on the *very first* months of life offers additional implications for the consideration of treatment approaches.

Finally, we have three remaining questions for Dr. Schore:

1. Given his emphasis on the baby's brain's growth occurring in the context of a "positive affective relationship," or that attachment happens with the caregiver who "expands opportunities for positive affect and minimizes negative affect," how would his model account for the data showing attachments to sadistic objects as well as explain the differing effects on mind-brain and emotional development of aversive stimulation versus understimulation?

2. It seems inconsistent to hypothesize that minds are "psychobiologically" differently equipped according to gender, as Dr. Schore does, when his work amply

demonstrates the *social* construction of this psychobiology. If the brain's emotional regulation is built from
the mother–child interaction, couldn't the divergent
ways mothers interact with boys and girls contribute
to the variation in the male and female brain?

3. Why use the term *imprinting* when the neurological
   data, including Eisenberg's (1995) work, point to a
   construction of the brain in terms of neurological
   connections and neural circuitry? The infant's brain is
   literally getting made up from the mother–child interaction, so why use a term such as imprinting, which
   suggests a learning process between two relatively
   formed minds?

References

Bach, S. (1994). *The Language of Perversion and the Language of Love*. Northvale, NJ: Jason Aronson.

Brazelton, T. B. (1992). *Touchpoints: The Essential Reference: Your Child's Emotional and Behavioral Development*. Reading, MA: Addison-Wesley.

Brazelton, T. B., and Cramer, B. G. (1990). *The Earliest Relationship: Parents, Infants, and the Drama of Early Attachment*. Reading, MA: Addison-Wesley.

Condon, W., and Sander, L. (1975). Synchrony demonstrated between movements of the neonate and adult speech. *Child Development* 45:456–462.

Eisenberg, L. (1995). The social construction of the human brain. *American Journal of Psychiatry* 152:1563–1575.

Ellman, S. J. (1991). *Freud's Technique Papers: A Contemporary Perspective.* Northvale, NJ: Jason Aronson.

—— (1992). Psychoanalytic theory, dream formation, and REM sleep. In *Interface of Psychoanalysis and Psychology,* ed. J. W. Barron, M. N. Eagle, and D. L. Wolitzky, pp. 357–374. Washington, DC: American Psychological Association.

—— (1995). The Edmund Weil Memorial Lecture: The plight of post-Freudian theories: Can they stand without Freudian scaffolding? The Institute for Psychoanalytic Training and Research, New York, May.

—— (1996). What is unique about Freudian technique? Lecture presented at the New York University Freudian Psychoanalysis Today Conference, November.

Ellman, S. J., Spielman, A. J., and Lipschutz-Brach, L. (1991). REM deprivation update. In *The Mind in Sleep,* ed. S. J. Ellman and J. S. Antrobus, pp. 369–376. New York: Wiley.

Ellman, S. J., and Weinstein, L. (1991). REM sleep and dream formation: a theoretical integration. In *The Mind in Sleep,* ed. S. J. Ellman and J. S. Antrobus, pp. 466–488. New York: Wiley.

Freud, S. (1915). Instincts and their vicissitudes. *Standard Edition* 14:117–140.

—— (1923). The ego and the id. *Standard Edition* 19:12–59.

Jacobson, E. (1964). *The Self and the Object World.* Madison, CT: International Universities Press.

Jouvet, M. (1967). Neurophysiology of the status of sleep. *Psychological Reviews* 47:117–177.

Lester, B. M., and Zeskind, P. S. (1982). A biobehavioral perspective on crying in early infancy. In *Theory and Research in Behavioral Pediatrics,* vol. 1, ed. H. Fitzgerald et al. New York: Plenum.

MacFarlane, A. (1975). Olfaction in the development of social preferences in the human neonate. *Parent–Infant Interaction*. CIBA Foundation Symposium 33. New York and Amsterdam: Elsevier.

Spielman, A. J., Ellman, S. J., and Steiner, S. S. (1973). The effects of varying amounts of intracranial self-stimulation on the normal sleep cycle of the rat. *Psychophysiology* 11:171–174.

Steiner, S. S., and Ellman, S. J. (1972). Relation between REM sleep and intracranial self stimulation. *Science* 177:1122–1124.

Steingart, I. (1995). *A Thing Apart*. Northvale, NJ: Jason Aronson.

Winnicott, D. W. (1956). Primary maternal preoccupation. In *Through Paediatrics to Psycho-Analysis*. New York: Basic Books, 1975.

——— (1960). The theory of the parent–infant relationship. In *The Maturational Processes and the Facilitating Environment: Studies in the Theory of Emotional Development*. Madison, CT: International Universities Press, 1991.

3

The Contribution of Self- and Mutual Regulation to Therapeutic Action: A Case Illustration

Frank M. Lachmann and Beatrice Beebe

Empirical infant research can expand our understanding of therapeutic action in adult treatment (Beebe and Lachmann 1994, Lachmann and Beebe 1992a, 1996). The concepts of self- and mutual regulation, derived from a systems approach to the study of infant–caregiver interaction, permit a more detailed view of patient–analyst interaction and the processes of analytic change. A case is used to illustrate the origins and transformations of a patient's psychopathology. In this case, chronically mismatched interactive regulations led to premature, drastic self-regulations. We trace the interactive processes of therapeutic action.

We have previously offered three organizing principles derived from infant research to describe how interactions are regulated, represented, and begin to be internalized in the first year of life (Beebe and Lachmann 1994, Lachmann and Beebe 1996). These principles are ongoing regulations, disruption and repair of ongoing regulations, and heightened affective moments. They constitute hypotheses about how social interactions become patterned and salient. Although infant research has been construed as relevant to adult treatment in many ways (Emde 1988, Homer 1985, Osofsky 1992, Sander 1985, Seligman 1994, Soref 1992, Stern 1985, 1995), in this chapter we focus on the principle we consider to be overarching, ongoing self- and mutual regulations. Through metaphor and analogy, we apply this perspective to analyst–patient interaction, verbal and nonverbal, in one case.

We use a theory of the interactive organization of experience that is based on a dyadic systems view (Beebe et al. 1992,

Lachmann and Beebe in press, Samaroff 1983, Sander 1977, 1983). In this perspective, organization is an emergent property of the dyadic system and a property of the individual. Thus this view integrates the simultaneous influences of mutual and self-regulation. Mutual regulation refers to a model in which both partners actively contribute to the regulation of the exchange, although not necessarily in equal measure or in like manner. Self-regulation refers to self-comfort and the capacity to regulate one's states of arousal and organize one's behavior in predictable ways (Beebe and Lachmann 1994).

The necessity for integrating both self- and mutual regulation in early development argues for their integration in a psychoanalytic theory of adult treatment as well (Beebe and Lachmann 1994, Lachmann and Beebe 1989, 1992a). Mother and infant, and analyst and patient, jointly construct patterns of social relatedness. These patterns guide the management of attention, participation in dialogue, and affect sharing. Each partner influences the process through his or her own self-regulatory range and style, and through specific contributions to the pattern of interaction (Lachmann and Beebe 1996).

The organization of experience is the property of the interactive system and of the individual (Sander 1977, 1985). Experience is organized as predictable; coordinated rhythms, tempos, sequences, affective intensities, postures, greetings, and separations unfold. At the same time, self-regulatory styles emerge. Characteristic and expected adaptive and non-adaptive patterns of mutual and self-regulation are thus constructed (Beebe et al. 1993, Beebe and Lachmann 1994, Tronick 1989).

This model of development derived from infant research cannot, of course, be directly translated into the adult psychoanalytic situation. In adults, the capacity for symbolization and the subjective elaboration of experience in the form of fantasies, wishes, and defenses further modifies the organization and representation of interactive patterns. However, what makes this model appealing for adult treatment is that it makes no assumptions about the dynamic content of adult experience. It focuses entirely on the process of interactive regulation (Lachmann and Beebe 1996).

The person's capacity to respond and be socially engaged depends not only on the nature of the partner's input, and on the nature of the person's responsivity, but also on the person's regulation of his internal state (Beebe and Lachmann 1994). State is used broadly to refer to affect, arousal, and its symbolic elaboration. From infancy on, people differ in the crucial capacity to modulate arousal, shift state, and tolerate and use stimulation to organize behavior in predictable ways (Als and Brazelton 1981, Korner and Grobstein 1977).

Specific failures in self-regulation affect the quality of mutual regulation. For example, infants with specific self-regulatory difficulties may place undue burdens on the responsivity of their partners. Whether derived from variations in individual endowment or failures in mutual regulation, difficulties in self-regulation affect the quality of engagement.

Likewise, specific failures in mutual regulation compromise self-regulation. For example, as in the case to be discussed, affect regulation (Socarides and Stolorow 1984/85), anxiety, tension, and aloneness (Adler and Buie 1979) may then be relegated to solitary measures, without a sense of support. Expectations of being abandoned when one is vulnerable may arise. Rather than an expanding self-reliance and self-sufficiency, aversions, anxieties about relationships, or self-protective depersonalization may ensue (Beebe and Lachmann 1994).

Our integration is not designed to supplant dynamic formulations. Instead, it can provide the analyst a more differentiated view of the regulation of interactions and the organization of experience that goes beyond interpretation. Numerous well-established psychoanalytic concepts already cover some of the same terrain as self- and mutual regulation. For example, ongoing regulations have been subsumed within discussions of patterns of transference and countertransference (Lachmann and Beebe 1992b), the "holding environment" (Winnicott 1965), and the "background of safety" (Sandler 1960).

In the treatment of the case to follow, we emphasize the analyst's and patient's mutually regulated interactions of affect, mood, arousal, and rhythm. Nonverbal interactions at the microlevel of rhythm matching, modulation of vocal contour, pausing, postural matching, and gaze regulation are usually not given adequate recognition in the treatment process (Beebe 1993). We describe their powerful moment-by-moment impact on the joint construction of the psychoanalytic relationship, and we track the alterations in the self-regulatory ranges of both patient and analyst.

Attention to self- and mutual regulation helps the analyst contact difficult-to-reach patients where the critical cues go far beyond the usual verbal exchange. This perspective also expli-cates the interactive construction of selfobject experiences.

We are not proposing a new technique, nor are we argu-ing for decreased attention to dynamic issues in treatment. Instead, we reverse the figure-ground perspective customarily used in describing analytic treatment. We place the mutually regulated nonverbal exchanges into the foreground and dynamic conceptualizations into the background.

The nonverbal interactions on which we focus have been included among noninterpretive analytic behaviors (e.g., Fer-enczi 1929, Freud 1909, Lindon 1994). These interventions have been made when words were considered inadequate to retain a therapeutic connection with certain patients. However, we hold that nonverbal interactions, like noninterpretive actions, do constitute interpretations, although not packaged in the customary form. Their intent is to provide a primary contri-bution to the patient's expectation of mutuality and being understood. They provide access to a patient who is in a state not accessible to more usual forms of therapeutic dialogue.

Clinical Illustration

Karen (treated by F. L.) began psychoanalytic therapy after her first suicide attempt. When her then-boyfriend flirted with another woman, she took all the pills in her med-icine cabinet. Like an automaton, she watched her actions "from a bird's eye view from a corner of the room." This detached, depersonalized state felt to her as though she were "behind a pane of glass."

At times Karen would become aware that she had lost several hours and would find herself in a different part of town, with no memory of how she got there. In one of these states, toward the end of her first year of therapy, Karen made a second, similar suicide attempt.

Karen is in her sixth year of treatment on a three-session-per-week basis. When she began therapy, she was 27 years old. Her life had been declining since her graduation from high school. At that time her parents literally dragged her from her room to deposit her at an out-of-town college. She succeeded in remaining there for four years, excelling in some of her courses, and at the same time barely passing others. After graduation, she gained admission to a drama school in England and studied there for a year.

Upon her return to the United States, Karen continued to study acting and attempted to find work as an actress. She suffered from a severe sleep disturbance. Unable to sleep during the night, she tended to fall asleep in the early morning hours. When awake, she could not mobilize herself. She thus missed many auditions, missed call-backs when she did have a successful audition, and failed to appear for acting jobs when she did get hired.

Karen found it difficult to speak to me. She felt she had nothing to say and began sessions by asking, somewhat mechanically, "What shall we talk about?" Initially, I responded by summarizing previous sessions, for example: "We spoke about how messy and dingy your apartment is." She might then speak about an apparently unrelated topic, for example, encounters with various acquaintances. I

thought these experiences left her with feelings of abandonment, disappointment, a sense of exploitation, or regret
about her withdrawn manner.

Privately, I came to understand Karen's opening question
as an attempt to orient herself and determine whether or not
a connection could be established with me. I believed that
waiting for her to begin, or throwing her question back,
would have failed to recognize her tentative attempt to reach
out to me. Overtly Karen hardly acknowledged my presence,
although she did ask, "What shall 'we' talk about?"

Gradually I came to appreciate that Karen dreaded con-
versing with me. She anticipated that she would have difficult
feelings she would have to regulate on her own. This expectation suggests an imbalance between mutual and self-regulation in her development. With the expectation of chronic
misregulation, a preoccupation with self-regulation and the
management of negative affect ensues (Gianino and Tronick
1988, Tronick 1989). For Karen, despite her preoccupation
with self-regulation, the nature of her self-regulatory efforts
were severely impaired. Her sleep disturbance, listless state,
and lack of "desire" were evidence of this impaired regulation
of affect and arousal.

What self-regulatory range did Karen bring to treatment? A
narrow range of tolerable affect, arousal, and engagement; an
immobile face; a tendency to space out; and massive efforts to
dampen down her reactivity to all stimulation. In sessions she
looked out of my office window as she sat in her chair with her
coat on. When she did look at me, it was a sideways glance.
Interactions with Karen were dominated by her withdrawal.

Slowly Karen began to reveal her experiences in her family. From the age of 5 on, she was awakened during the night by her parents' fights. She could hear their shouts through the walls of her bedroom. Her mother would accuse her father of staying out at night with other women, and her father would, at times, be physically abusive toward her mother. The fights terrified her, especially after her mother asked her who she would want to go with if she and her father separated. She remembered not wanting to go with either one. They never did separate.

Karen was born shortly after her parents' graduation from high school. Neither parent hid from Karen that their future plans were scuttled by their marriage and her birth. In reaction, Karen began to pray at night. By the time she was 7, she had made a deal with God. If He would stop her parents from fighting, she would give up her life. From that time forward she was preoccupied with suicide and suffered from severe, persistent sleep disturbances.

In high school, Karen frequently cut classes because she could not tolerate the noise made by other students. Sometimes she would get as far as the classroom door, stand outside, and be unable to enter. She would then go home and spend the day studying alone. She did appear for examinations, on which she did very well. During afternoons and evenings she worked as a cashier at a local shopping mall. In fact, she worked continuously until the end of her high school years. Since then and until her treatment was well under way, she had not held a job.

By the age of 17, Karen had formed an intimate sustaining relationship with a fellow student, Brian. He was her best friend and confidant. When he unexpectedly died of a brain tumor, she was despondent. Her parents insisted that she get over her loss quickly. They could not recognize that Karen had lost the one person to whom she felt close and whom she trusted. She retreated to her room, felt "without desires," and was increasingly aimless. This was not the first time she experienced these states. On previous occasions they were more or less transient. At this point they crystallized as a recurring dominant state.

When Karen began therapy, ten years later, these states were still prominent and affected the nature of our interaction. Though we sat facing each other, Karen looked away from me. Her face was immobile. Her voice had no contouring. She kept her coat on. Even when I spoke to her, she did not look at me. Her self-constriction powerfully affected me. I felt reluctant to jar her precariously maintained presentation.

I responded to her constriction by partially constricting myself. I allowed myself to be influenced by her rhythm. I narrowed my own expansiveness to more closely match the limits imposed by her own narrow affective range. I did look at her continuously, but I kept my voice even and soft. In my initial comments I remained within the limits of the concrete details that she offered. I thus altered the regulation of my own arousal, keeping it low and limiting my customary expansiveness. She was effective in communicating her distress, and I was able to respond by providing her with a range of stimulation that more closely matched the limited level of arousal she could tolerate. However, as I restricted my own expressiveness, at times I became fidgety and squirmy. She seemed oblivious to my moments of discomfort.

Gradually her tolerance for affective arousal increased, and I could become more expansive. She was able to talk about affectively more difficult material. Her voice remained soft but with more contouring. She spoke about social relationships and acting auditions that raised the specter of competition. But she felt that she had no right to live. She withdrew from these situations lest she draw attention to herself. Initially these explorations had little effect on her life. Our dialogue, however, did increase the affective range that she could tolerate in the sessions.

During the first two years of the treatment, Karen moved from descriptions of her environment, the inanimate world, to descriptions of interpersonal relationships and explorations of her subjective states. At the same time I also shifted from summaries of the previous sessions to elaborations of her feelings and reactions. Sometimes anticipating her formulaic opening, "What shall we talk about?" I began sessions by asking her how she was feeling. Sometimes, before verbally responding, her right upper lip would twitch and constrict briefly, or her leg would jiggle rapidly. I came to understand these signals as an indication that she was tense and had been feeling moody, depressed, or without energy since our last meeting. We focused on her specific reactions and tried to find a context for them. I detailed nuances of feelings and moods such as annoyance, rebuff, eagerness, enthusiasm, or disappointment. I told her that it seemed to me that she experienced many emotions as though they were annoying intrusions. After some time, I was able to add descriptions of Karen as "considering," "hoping," "planning," or "expecting." That is, I distinguished among categories of affect and time, and acknowledged her authorship of her experience. I kept apace with Karen's gradually more personalized communications. The extent of her visible discomfort waxed and waned. However, she did appear more comfortable about accessing, revealing, and understanding her subjective life in both its reactive and proactive aspects.

Although I was not unaware of restricting and monitoring my responses to Karen, I was not following a premeditated plan of nonverbal treatment. Qualities of nonverbal communication, such as vocal rhythm, pitch, contour, and the level of arousal are usually out of awareness, but we are able to bring them into focused attention. It was mostly in retrospect that I became aware of the salient role played by these nonverbal, mutually regulated interactions and their effect on Karen's and my self-regulation. I assumed that through these interactions, Karen felt some sense of validation leading to the tentative engagement of a selfobject tie.

To enable Karen to maintain the fragile developing selfobject tie, I did not make explicit the possibility that Karen refound aspects of her experience with Brian in the therapeutic relationship. To do so could have increased her self-consciousness and her propensity toward overstimulating anxiety. She would have needed to protect herself against a repetition of another attachment and loss-abandonment sequence.

Instead, I recognized Karen's affective reactions and her dread of retraumatization in her current relationships. We

translated her associations, symptoms, nightmares, enactments, and hallucinations into more direct statements about
herself and her experience. We discussed the vagueness that
characterized much of her life as indicating what she was afraid
to perceive, feel, believe, wish, imagine, or remember. My
comments were directed toward recognizing her attempts at
self-definition. In this way her previous, almost exclusive reliance on drastic self-regulation through withdrawal, depersonalization, and derealization began to shift. Her sense of agency
increased. For example, as her dread of retraumatization in
new situations diminished, she showed a wider variation of
facial expressions. Occasionally she smiled. Furthermore, she
registered for, attended, and participated in some classes.

In the course of the first two years of treatment, Karen
recalled her parents' fights and was increasingly able to
describe other painful events. However, until we explored
them, these memories had been retained as unconnected
experiences. They correlated with Karen's sense of fragmentation. In making connections among these memories, her
feelings, and their current relevance, I continually depicted
Karen as living and having lived a life with temporal, affective, and cognitive dimensions. I described the events to her,
with some slightly increased affective elaborations. Increasingly she could tolerate my amplification of her affect.

These memories spanned the fifth to eighth years of her
life. In the earliest event we pieced together, Karen was
required to mail a letter for her mother at the post office.
Karen recalled feeling abject terror at having to walk past
some derelicts to mail the letter.

Karen recalled that she could not leave her mother's side. Nor could she explain to her mother why she was so afraid. her mother encouraged her to mail the letter by telling her how big and grown up she was. We came to understand that walking alone past the derelicts meant to Karen that she would be showing her mother that she was "growing up." She would then be telling her mother that she was able to fend for herself. Based on the fights Karen had overheard, I said that her fear of revealing her "growth" grew out of her belief that her mother wanted to "dump" her and, in fact, could not wait to do so. Showing independence would result in imminent loss of support. Karen responded with silent tears. She rarely responded to such reconstructions directly. However, new recollections subsequently did emerge.

Another memory concerned a visit to a department store where Karen had one of her first asthmatic attacks. She tried to tell her mother that she could barely breathe. She could not keep up with her as they rushed from one department to another. Her mother told her that the shopping was important. Karen should not complain so much.

Karen recalled the department store visit as we were exploring how she neglected her health, especially her teeth, skin, and allergies. I commented that through her body and frequent upper respiratory illnesses, her complaints were given voice. There she retained an eloquent record of her feelings. The twitch of her upper lip and her foot jiggles also constituted such a silent record of her moods and feelings.

Karen had come to consider her physical state as unimportant. She could now imagine that she must have felt hurt by her mother's dismissive and neglectful behavior. She had been anxious about her breathing difficulty, but most of all, fearful of evoking her mother's disapproval by complaining. Her solution in the department store had been to redouble her efforts to stay close to her mother, attempt to dampen her own arousal, and hope that the ordeal would soon be over.

The third memory involved Karen's sleep difficulty, which continued even after the parental fights ceased. To cure her sleep difficulty she was confined to her room. While exploring this symptom, Karen reported that she was currently feeling a sudden urge to travel to Iceland. She then recalled that she had attempted to deal with her sleep problem by putting herself to sleep in an empty bathtub. There she could fall asleep because "it was quiet." But, most important, the "whiteness" and the hardness of the tub felt so good. It afforded her a sense of security and protection. However, when her parents discovered her, she was sent to her room immediately after supper so that she could try to get to sleep early.

Obediently Karen remained in her room. She watched the outside world from behind her window. She felt excluded and frustrated and began to draw on the walls of her room. These creative, exploratory, assertive efforts were quickly punished. She was given a bottle of cleaning fluid and told to undo the damage. Later, however, she was presented with a paint set, but she felt too resentful toward her parents to use it. The paint box was never opened. Some years later, in a similar vein, she was given a chemistry set because she showed some interest and ability in her science classes. However, aside from being presented with the paints and chemicals, no one in her family took any interest in her. No one inquired about the two sets. Neither was ever used. Karen acknowledged that she would have liked to use the paint and chemistry sets but could not bring herself to do so.

Karen clung to her room until she was dragged by her parents to college. Her room was her refuge and she felt protected in it, even though alone. Her banishment and "voluntary confinement" to her room paralleled my sense of her inaccessibility in the therapeutic relationship. I made this connection with the expectation that Karen's inaccessibility could be explained and thereby diminished more directly. She thereupon dreamed of a "barren countryside." In association she recalled hallucinatory experiences. When she was confined to her room and would look out of her window, she sometimes saw cars go by without drivers. We discussed the barren countryside dream, a self-state dream, as conveying her sense of barrenness. We connected it to the aimlessness depicted in her hallucination. I told her that she depicted herself as living in a world in which no one was at the wheel. Perhaps she longed for someone in her family to take an interest in her and assume some control. I also inquired whether she might be feeling this way currently in her treatment. Such heavy-handed transference queries never yielded much. On further reflection, however, I felt that my expectations of her exceeded a level of functioning that she could tolerate. Through her dream and her recollection of the hallucinations, she was reminding me of her still-depleted state, her barrenness. She was letting me know that I should not rush ahead of her (as her mother had in the department store) and lose sight of the severity of her difficulties. Her dread of being dumped at the first signs of growth still prevailed.

The three memories of being at the post office, in the department store, and confined to her room constitute a series of model scenes (Lachmann and Lichtenberg 1992, Lichtenberg et al. 1992). Each one shaped segments of our interactions, though the "confinement" theme dominated the others.

The model scene of standing in her room and looking out at life through her window gathered together a number of previous salient issues and shaped subsequent experience. In her room, she was protected from the injurious expectation of her family and the "noise" and potential exploitativeness of her peers. In her room she did not need to fear that she might "blow others away" or become an object of envy. She also avoided the danger of feeling helpless, frustrated, and out of control. Through her continuous self-sacrifice, she maintained a firm grip on her parents' tie to her and her claim on them.

The mother–daughter relationship, encapsulated by the model scenes, depicted patterns of mutual regulation that tilted Karen toward solitary self-regulation. At first she did attempt to regulate heightened, painful affect states, such as her terror at the post office and in the department store, by trying to elicit her mother's participation. She pleaded with her mother at the post office and clung to her in the department store. Feeling ignored, she expected that independent steps would lead to abandonment. Therefore, heroic efforts at self-regulation were undertaken. In essence, drastic self-regulation attempts substituted for a balanced integration between self- and mutual regulation (Tronick 1989).

By the time Karen was confined to her room, she had come to tolerate her aloneness and restrict her activity. Her efforts to engage her family had all but ceased. Her physical symptoms and hallucinations increased her withdrawal. She felt ineffective in engaging her parents and confined herself to altering and influencing her physical and subjective states. Drawing on her walls served as a desperate signal for attention. But her refusal to touch the paint and chemistry sets suggested that her withdrawal contained a significant degree of self-sacrifice and defiance. Karen could not risk putting herself in the position of expecting recognition from her parents and being disappointed. To avoid this danger she kept her creative and intellectual interests to herself. She lived out her grim, unconscious belief (Weiss and Sampson 1986) that she had "no right to a life" (Modell 1984). Her suicidal preoccupation, her neglect of her physical well-being, her propensity to disregard physical illness, her social withdrawal and minimum functioning in life, and her retreat from attention no matter how much she desired it all converged in her conviction that her parents' life (and hence the world) would have been better off had she not been born.

Her relationship with Brian in late adolescence provided a notable exception to these convictions. Through her relationship with him, Karen had retained some hope for a reciprocal connection and sensual and sexual responsivity. A significant sector of her life had been left relatively intact. However, overall her development remained constricted.

Karen developed physical symptoms in lieu of accusations and complaints, withdrew from people, and found solace in the bathtub. The bathtub differed as a solution from the others in that she created her own protected environment. Her confinement to her room thus became an enforced exclusion from her family. Yet it provided some protection against the overstimulating parental quarrels and her parents' obliviousness to her needs.

During the treatment the self-protective aspects of Karen's bathtub experience appeared transformed as a visit to Iceland. Karen began to acknowledge the talents that were unrecognized by her family. Her creativity remained sequestered in her private world with considerable ambivalence. For example, she studied acting but did not perform.

Convinced that she had been a burden to her parents and the source of their difficulties, Karen considered her solitary confinement to be justified. In refusing to make use of the resources that her parents gave her, she found a self-defeating but nevertheless modest triumph. In the confines of her imagination, creative elaborations of her experience continued in silence and in private. These could be accessed in the course of her treatment and became the imagery of her poetry.

Karen had written poems at various times in her life. During her second year of treatment she turned to writing poetry in a more determined way. A poem she brought to a session was dedicated to the memory of Brian. In it she depicted her loneliness and her alienation from her family. She ended with a plea to Brian: "Run after me but never let me go."

To use therapy, for which her parents paid, meant to Karen that she had to surrender her defiance and capitulate to them. She had not used the paint box, the chemistry set, or the acting classes. Why surrender now? It became clear then, why, during the first two years of treatment, she continued to indicate that she had made no progress and that she was as depressed as ever. Based on Karen's failure to work and earn money, her parents echoed her feeling that she had not made any progress in therapy. I asked Karen what would happen if her parents were to stop their financial support of her. She said, "I would probably be dead, now." It was as though Karen was giving her parents another chance to decide, Do you want me to live or not? Since the financial support included payment for her treatment, I also understood her remark to allude to her need for therapy and its importance to her.

During the first two years of treatment, Karen usually came to the sessions encased in the room in which she was isolated by herself and her family. She either stayed in her room literally by not coming to sessions, or figuratively through her communicative difficulties. Often sessions felt to me as though we were meeting for the first time. She never made any reference to what had gone on in a previous session, so I did. She did appear to be moved by some of my affect-laden descriptions of her experience. When she was moved, tears would roll down her cheeks. She could not usually say why she was crying.

During these first two years of treatment, Karen missed at least one of her three weekly appointments and arrived late for the other two. Missing sessions or arriving late increased her sense of failure. When I gently inquired about this pattern, she told me that it was an achievement that she could get herself to the sessions as often as she did.

During these first two years, I referred Karen to a psychopharmacologist, but she did not take the medication with any regularity. Fortunately, the unused medication was not at hand when she made her second suicide attempt. During the first two years, Karen also had two abortions to which she reacted with increased depression. Twice during the second year of her treatment I met with Karen and her parents, but they could not grasp the severity of Karen's difficulties. I felt that she was still a high suicide risk.

After the second suicide attempt, waiting to see whether or not Karen would arrive for her appointments became very anxiety arousing for me. I felt that without some more active intervention on my part, her depersonalized state and the suicidal potential would continue. I needed more reliable and intensive contact with her. I needed to feel less worried and that I had a chance of reaching her.

Thus, at the beginning of her third year of treatment, I decided to telephone Karen about two hours before every appointment. I reminded her of the time of our meeting and told her that I looked forward to seeing her. Within about three to four months, Karen no longer missed sessions.

Karen had engaged me sufficiently that I had begun to feel desperate. When I had decided to call her, I was not conscious of her plea in the poem, "Run after me but never let me go." However, I was evidently responding to it. In retrospect, my enactment exactly matched the presymbolic quality of many of Karen's communications. We may ask whether her long-standing, continuously reinforced conviction that she was fundamentally unwanted would have budged in the face of verbal interventions and explanations alone. Could attuned understanding have better facilitated the therapeutic process? Were my calls an extension of empathic immersion in her subjectivity? Or was I requiring Karen to connect with me on my terms and at her expense?

Though valid, these questions imply that my self-regulation and my role in the mutual regulation could have been reduced or eliminated. Although a dramatic departure from customary analytic work, the telephone calls emerged out of a mutual regulation in which my capacity to tolerate anxiety had reached a limit. Furthermore, my enactment concretely mad the following interpretation to Karen: you are wanted. We hold that this enactment was a critical part of the regulatory process and therapeutic action in this case.

Despite Karen's detachment, her responsivity to some of my comments did evoke an intense engagement on my part. Her dramatic response to my calls about coming regularly to sessions indicated how profoundly she could be influenced by me. Her response exactly matched what I needed to feel, so that the treatment—and she—had a chance. Not only was her sense of efficacy promoted as I altered and restricted my responsivity, but my sense of efficacy was promoted as she expanded her responsivity. Thus a complex and intricately matched mutual regulation took place.

By the end of the third year, the gradually firming selfobject tie made suicide less likely and diminished her depression. She had to admit that she had not felt so well in many years. She even volunteered that she did not think she could ever make another suicide attempt.

During the first year of my calls, if she was still asleep, her answering machine would pick up and I would leave a message. As Karen became less depressed and felt more energetic, she often left her house before my call. She would then come to the sessions without a reminder and would receive my message only upon arriving at her home in the evening. On several occasions I asked how she felt about my telephone calls. She told me that it was "OK" for me to call. I understood her "OK" as her only way of saying that she wanted the calls. She could not acknowledge that she needed them. With her "OK" she gave me permission to call as if also reassuring me that I was not intruding. In this response it is apparent that Karen was still quite detached and protected herself in the privacy of her room.

By the beginning of the fourth year of treatment, Karen appeared more alive and accessible. The gradual establishment of a relatively reliable selfobject tie shifted Karen's self-regulatory capacities toward greater tolerance for affect and arousal. Although she remained rather constricted, she gained increasing access to her own experience and her own history. Past and current impressions gained expression in her writing.

In this fourth year Karen wanted to talk to her mother about the recent death of an acquaintance. It was an event that bore certain similarities to the death of Brian. Her mother suggested that they meet at a bar that had music to talk about this death. Karen then telephoned me. She had felt guilty about her actions at the last meeting with this acquaintance and she was now also disappointed and furious with her mother for fending her off. Thus, in spite of her continuing state of detachment, she was able to use our tie to restore herself in this crisis.

In her fifth year of treatment, Karen's interest and talent as a writer enabled her to enroll in a graduate program, attend classes, and submit assignments. Through her visit to Iceland she succeeded in having some poetry published. Though she still sought relationships with charismatic men who were unstable and irresponsible, she was no longer so compliant and dependent. She practiced safe sex.

Karen's conviction that she would cause fewer problems for her family by shutting herself away remained a dominant theme. In fact, it received continuous confirmation when she visited her parents. They did not seem to be aware of her widening range of affect and capacity. She was told by them not to come into the living room when they were entertaining friends because her depressing and uncommunicative manner put a pall on the company.

Karen is emerging as an adventuresome, foolhardy, overly trusting, resourceful, funny, and still somewhat self-sabotaging person. In her own succinct way she summarized her gain in her treatment: "I used to not be able to talk to people. Now I can talk to people."

Discussion

Karen's lifelong experiences of rebuff led to a premature reliance on drastic and restricting self-regulatory measures such as avoidance, depersonalization, derealization, and dampening of her own affect. Designed to avoid retraumatization, these measures only partially protected her. She maintained a precarious balance between self-expression and self-annihilation.

In Karen's development, sounds had become shattering noises obstructing emotional contact. Vision had become a remote sense. She felt as though she were looking at herself and her experiences from a distance. Breathing, sleeping, and spatial orientation were impaired. Sensual-affective experiences were overarousing, emerged as physical symptoms and disruptive imagery such as hallucinations and nightmares. Unable to regulate affect states on her own, she avoided emotionally arousing and thus potentially disruptive experiences.

The relationship with Brian revived Karen's expectations of being affectively validated and part of a dyad. It rekindled her expectation that she could trust her feelings, be included in someone's internal life, and form a bond. We assume that the tie to Brian reengaged an earlier, precarious selfobject tie to her parents. With the death of Brian, Karen was traumatized (Lachmann and Beebe in press). Not only did she lose the only person to whom she felt connected, but her parents also failed to validate her profound devastation. Thus, selfobject ties were irreparably disrupted.

The treatment of Karen illustrates the role of mutual and self-regulation in the therapeutic establishment of the selfobject tie. Karen's fears of retraumatization were investigated, and her feelings were labeled, differentiated, and affirmed. As with Brian, she feared that attachment would lead to loss. Furthermore, her restricted self-regulatory range interfered with her ability to tolerate the excitement and hope generated by the expectation of being accepted, understood, and included in a bond. These difficulties pervaded her friendships, classes, and work possibilities as well as her treatment.

Karen's immobile face, flat voice, sitting with her coat on, not looking, and having nothing to say required extraordinary measures. To reach her, the therapist had to restrict the range of affect and activity so that Karen's level of arousal remained tolerable to her. Speaking in a soft, even voice and slowing the rhythm increased Karen's tolerance for arousal. She began to talk about her life with a voice and face that were more alive. The therapist was able to expand the level of his own arousal and address her fears of retraumatization. In turn, Karen was less withdrawn. Fragments of her history emerged, from which three model scenes could be constructed. This increasing coherence led to Karen's ability to report a dream and a hallucinatory association of "cars without drivers." The therapist could then interpret her inaccessibility and her world where no one was at the wheel.

Although Karen was able to acknowledge that she would be dead without this therapy, her two abortions, extensive depersonalizations, second suicide attempt, frequent lateness, and continuously missed appointments led the therapist to make a dramatic intervention by telephoning her before every appointment. The fact that Karen was able to respond equally dramatically by coming regularly enabled the therapist to feel that he could continue to work with her. Karen was able to experience her own influence on the therapist's activity, and she could experience her therapist influencing her level of arousal. For both patient and therapist, self-regulation was altered through these mutual regulations. Thus, extensive work on Karen's depersonalized state and efforts to reregulate both her and her therapist had set the stage sufficiently well that the telephone calls could make a dramatic impact.

We have focussed on the nonverbal dimension in order to illustrate the contribution of mutual and self-regulation to therapeutic action. When the treatment began, solitary self-regulation was Karen's main method of survival, and it was failing. The treatment attempted to open up her self-regulation so that it could be included in a dialogue. Instead of viewing analyst and patient as two isolated entities, each sending the other discrete communications, we have illustrated a view of the treatment relationship as a system (Beebe et al. 1993). This theory of interaction specifies how each person is affected both by his or her own behavior (self-regulation) and by the behavior of the partner (interactive regulation) on a continuous moment-by-moment basis (Beebe 1993, Beebe and Lachmann 1988, 1994).

References

Adler, G., and Buie, D. (1979). Aloneness and borderline psychopathology: the possible relevance of child development issues. *International Journal of Psycho-Analysis* 60:83–96.

Als, H., and Brazelton, T. B. (1981). A new model for assessing the behavioral organization in preterm and fullterm infants. *Journal of the American Academy of Child Psychiatry* 20:239–263.

Beebe, B. (1993). *A dyadic systems view of communication: contributions from infant research to adult treatment.* Presented at the 16th Annual Conference of the Psychology of the Self, Toronto.

Beebe, B., Jaffe, J., and Lachmann, F. M. (1992). A dyadic systems view of communication. In *Relational Views of Psychoanalysis*, ed. N. Skolnick and S. Warshaw, pp. 61–81. Hillsdale, NJ: Analytic Press.

Beebe, B., and Lachmann, F. M. (1988). The contributions of mother–infant mutual influence to the origins of self and object representations. *Psychoanalytic Psychology* 5:305–337.

——— (1994). Representation and internalization in infancy: three principles of salience. *Psychoanalytic Psychology* 11:127–165.

Emde, R. (1988). The prerepresentational self and its affective core. *Psychoanalytic Study of the Child* 36:165–192. New Haven, CT: Yale University Press.

Ferenczi, S. (1929). The principle of relaxation and neocatharsis. In *Final Contributions to the Problems and Methods of Psychoanalysis*, pp. 126–142. New York: Basic Books, 1955.

Freud, S. (1909). Notes upon a case of obsessional neurosis. *Standard Edition*, 10:153–318.

Gianino, A., and Tronick, E. (1988). The mutual regulation model: the infant's self and interactive regulation and coping and defensive capacities. In *Stress and Coping*, ed. T. Field, P. McCabe, and N. Schneiderman, pp. 47–58. Hillsdale, NJ: Lawrence Erlbaum.

Horner, A. (1985). The psychic life of the young infant: review and critique of the psychoanalytic concepts of symbiosis and infantile omnipotence. *American Journal of Orthopsychiatry* 55:324–344.

Korner, A., and Grobstein, R. (1977). Individual differences at birth. In *Infant Psychiatry*, ed. A. Rexford, L. Sander, and T. Shapiro, pp. 66–78. New Haven, CT: Yale University Press.

Lachmann, F. M., and Beebe, B. (1989). Oneness fantasies revisited. *Psychoanalytic Psychology* 6:137–149.

——— (1992a). Reformulations of early development and transference: implications for psychic structure. In *Interface of Psychoanalysis and Psychology*, ed. J. Barron, M. Eagle, and D. Wolitzky, pp. 133–153. Washington, DC: American Psychological Association.

——— (1992b). Representational and selfobject transferences: a developmental perspective. In *Progress in Self Psychology, vol. 8, New Therapeutic Visions*, ed. A. Goldberg, pp. 3–15. Hillsdale, NJ: Analytic Press.

——— (1996). Three principles of salience in the patient–analyst interaction. *Psychoanalytic Psychology* 13:1–22.

——— (in press). Trauma, interpretation, and self-state transformation. *Psychoanalysis and Contemporary Thought*.

Lachmann, F. M., and Lichtenberg, J. D. (1992). Model scenes: implications for psychoanalytic treatment. *Journal of the American Psychoanalytic Association* 40:117–137.

Lichtenberg, J. D., Lachmann, F. M., and Fosshage, J. (1992). *Self and Motivational Systems*. Hillsdale, NJ: Analytic Press.

Lindon, J. (1994). Gratification and provision in psychoanalysis: Should we get rid of the rule of abstinence? *Psychoanalytic Dialogue* 4:549–582.

Modell, A. (1984). *Psychoanalysis in a New Context*. New York: International Universities Press.

Osofsky, J. (1992). Affective development and early relationships: clinical implications. In *Interface of Psychoanalysis and Psychology*, ed. J. Barron, M. Eagle, and D. Wolitzky, pp. 233–244. Washington, DC: American Psychological Association.

Samaroff, A. (1983). Developmental systems: contexts and evolution. In *Mussen's Handbook of Child Psychology*, vol. 1, ed. J. Kessen, pp. 237–294. New York: Wiley.

Sander, L. (1977). The regulation of exchange in the infant–caretaker system and some aspects of the context–content relationship. In *Interaction, Conversation, and the Development of Language*, ed. M. Lewis, and L. Rosenblum, pp. 133–156. New York: Basic Books.

———— (1983). Polarity paradox, and the organizing process in development. In *Frontiers of Infant Psychiatry*, ed. J. D. Call, E. Galenson, and R. Tyson, pp. 315–327. New York: Basic Books.

———— (1985). Toward a logic of organization in psycho-biological development. In *Biologic Response Cycles: Clinical Implications*, ed. H. Klar and L. Siever, pp. 20–36. Washington, DC: American Psychiatric Press.

Sandler, J. (1960). The background of safety. *International Journal of Psycho-Analysis* 41:352–356.

Seligman, S. (1994). Applying psychoanalysis in an unconventional context: adapting infant–parent psychotherapy to a changing population. *Psychoanalytic Study of the Child* 49:481–500. New Haven, CT: Yale University Press.

Socarides, C., and Stolorow, R. (1984/1985). Affects and selfobjects. *The Annual of Psychoanalysis* 12/13:105–120. New York: International Universities Press.

Soref, A. (1992). The self, in and out of relatedness. *The Annual of Psychoanalysis* 20:25–48. New York: International Universities Press.

Stern, D. (1985). *The Interpersonal World of the Infant.* New York: Basic Books.

———— (1995). *The Motherhood Constellation.* New York: Basic Books.

Tronick, E. (1989). Emotions and emotional communication in infants. *American Psychology* 44(2):112–119.

Weiss, J., and Sampson, H. (1986). *The Psychoanalytic Process.* New York: Guilford.

Winnicott, D. W. (1965). *The Maturational Processes and the Facilitating Environment.* New York: International Universities Press.

4

Transference: A Self and Motivational Systems Perspective

James L. Fosshage and
Joseph D. Lichtenberg

To begin our discussion of transference, let us turn to the first reported session of Nancy's analysis, then in its second year.* Nancy addresses first her hurt and disappointment concerning the unavailability of her priest and, in a similar vein, her depression related to the weekend unavailability of her analyst. She mentions the stress of her exams, followed by her sexual "fantasies of being close and making love." She then reflects about the analyst to whom she is relating these matters. She hopes for a trusting relationship: "I trust I can confide in you and talk about it and feel better." Yet a different, frightening percept of the analyst also emerges, creating conflict: "On the news I heard about a psychiatrist who raped his patient. Here I am bringing all this explicit sex stuff in. What kind of person are you to want to hear about it, help me with it? Isn't there something perverse about it? What in all these cases gets out of control? The potential is there." The analyst wears the attributions of the "rapist" to further its elaboration in the "here and now" and inquires: "What you asked before, does it put me at risk for getting out of control? That I get stirred up as result of what you're talking about." Nancy reflectively replies that this is a possibility and that he (the analyst) has to do something with his

---

* A major portion of this paper appeared in a recent book entitled *The Clinical Exchange: Techniques Derived from Self and Motivational Systems*, co-authored by Joseph D. Lichtenberg, Frank N. Lachmann, and James L. Fosshage. The book centers around the analysis of Nancy. While space prevents the presentation of sessions in a verbatim format, we believe the clinical material of Nancy's analysis with J. L. will be amply clear to the reader.

pleasure and excitement. Reiterating her hope and fear, she declares, "If I entrust myself to you, I want to know you are able to deal with the stuff I bring up. I'm selfish—you could get out of control, or get deadened, unable to empathize." She then relates this frightening percept of the analyst to her experience of her father: "My relationship with you—I think of you as my dad. I was very close to him. I had to deal with feelings that would creep up. I have to deal with my own feelings about you, regardless of anything else. I'm aware of strongly stifling my curiosity about your life, desk, car—far removed from you personally. . . . You represent a verboten character."

Nancy then delineates themes derived from problematic thematic familial experiences that are currently active in the relationship with the analyst: Who was the seducer? Who was the seduced? Who was responsible? She says, "I'm not allowed to go around not fully dressed. Not being fully clothed all the way is attempting to seduce. So I got mad—the same stuff as Dad—I have to wear clothes to not disturb him. Nobody cares how it disturbs me! It's not fair. Why do I want to look and turn away? There's another class of verboten interests. I feel in all these cases I'm in the wrong and that's not right." Nancy presents a condensed but clear statement of the conflict that has been organized in relation to the analyst. She seeks to express sexual curiosity with its associated feelings and fantasies. Nancy's need for sexual excitement was consistently thwarted in her development, leading to years of a deadened genital anesthesia. As her sexual longings became manifest during the analysis, she was filled with shame and guilt. She feared that in expressing even curiosity, someone, either the man (father, brother, analyst) or herself, will get out of control and that, regardless of who will lose control, she is ultimately responsible for the seduction (what Ornstein [1974] has described as the dread of the repetition of the past). This leads into our discussion of the nature of transference and its two central aspects.

Patients enter a psychoanalysis with two broad groups of conscious and unconscious expectations. With dread, they expect to find in the treatment situation experiences that conform to prior problematic relationships. This expectation corresponds to the familiar model of transference as conflict derived from past conflict and trauma. Patients also expect their current analytic endeavor to provide growth-enhancing possibilities. Both of these divergent expectations in dialectic interplay influence the construction of the analytic relationship from the patient's perspective. This model of transference is similar in broad outline to Freud's unobjectionable positive transference and corresponds with Kohut's selfobject transferences when these are understood in a figure-ground relationship with the pathologic repetitive dimension (Stolorow et al. 1987) or representational configurations (Lachmann and Beebe 1992). We regard it necessary to explore both the transference constructions based on expectations of past problematic relationships and the expectations of new beginnings (Balint 1968, Ornstein 1974). We delineate the contents, affects, and conflicts associated with each, their sources in the present and past, and, of greatest importance, their interplay.

For Nancy, three abiding transference constructions stand out. In one configuration the analyst becomes the sexually threatening male and abandoning female, separately or in combination. In another view the analyst and analytic situation become a vehicle for soothing and calming, for attention focused on self-regulation, and for a thoughtful self-reflective approach to problems. Nancy created a third configuration by her distrust of the second view. A stance of the analyst that one moment she experienced as steady and thoughtful became diffident and indecisive, and a welcomed willingness of the analyst to engage with her in a consideration of sensual-sexual problems was then experienced as threatening and seductive without consummation or commitment.

Organization Model

In a significant departure from a drive displacement model (see Fosshage 1994), we, in concert with others (Fosshage 1994, Gill 1982, Hoffman 1983, 1991, Lachmann and Beebe 1992, Lichtenberg 1990, Stolorow et al. 1987), view transference experience as occurring within an interactive field and as variably co-determined by patient and analyst. The model we use assumes that all experience is organized (1) in conjunction with the context that is impacting perceptually, (2) in response to whatever motivational system is dominant, and (3) in accordance with expectations based on prior experience that have been generalized and presently activated. Transference refers to those particular experiences of analysands that focus on the analytic relationship.

Organization models of transference have emerged over the past decade and reflect a transition from positivistic to relativistic science (see Fosshage 1994, Gill 1982, 1983, 1994, Hoffman 1983, 1991, 1992, Hoffman and Gill 1988, Lachmann and Beebe 1992, Lichtenberg 1989, Stolorow and Lachmann 1984–1985, Wachtel 1980). While corresponding with what Hoffman and Gill refer to as the social-constructivist view, we designate it as an organization model (Fosshage 1994) to reflect the significance of the organization of experience as a core process for the developing sense of self. In its focus on the ongoing perceptual-affective-cognitive organization of experience, the model is anchored in recent developments in cognitive psychology (Bucci 1985), infant research (Stern 1985), and psychoanalytic developmental psychology (Lichtenberg 1983).

From our vantage point, the view that Nancy displaces and projects drive-organized distorted infantile object representations onto the analyst does not provide the encompassing view of the interaction between analyst and patient that we hold as definitive of the transference. Of course, thematic patterns established in her past exert a dynamic influence, but they are organized in the present in accord with an immediately perceived context. We do not view her disappointment in the priest as a displacement from her disappointment in the analyst, or from her parents, but as organized by the same thematic emotional experience that was triggered in her perception of each situation. Nancy's message is her disappointment in the absence of the other. In contrast to assuming Nancy's disappointment to be a displacement and shifting the focus to prior experience, we follow changes in Nancy's cognitive-affective state, occurrences in the analytic relationship, and thematic patterns that shape her perception of particular contexts. For example, Nancy's weekend separation from her analyst may have increased her vulnerability to the priest's lack of attentiveness and increased her susceptibility to disappointment. To view her emotional response as a displacement would depreciate the authenticity of her disappointment with the priest as an event with its own meanings, motives, and nuances to be explored. Furthermore, Nancy's desire to discuss her sexual feelings and fantasies triggers an image of the analyst that contrasts with her disappointment in the priest. She conceives of the analyst as a hoped-for and trustworthy man who is able to manage her and his own sexual feelings. She also conceives of him as frightening. He could become sexually aroused and lose control, or he has protected himself from sexual arousal by deadening his feelings, as she had for most of her life.

To view perceptions of the analyst as ongoing patterns that emerge, organize, and construct the analytic relationship compels the analyst to be alert to his and the patient's variable contributions to the triggering of these perceptions. Consequently, the analyst is drawn to live more fully in the here and now of the shared transference experience. Nancy's analyst, for example, engaged and explored the here and now of Nancy's transferential experience by inquiring, "What you asked before, does it put me at risk for getting out of control?" We propose that the exploration of the patient's experiential themes facilitates psychological reorganization through expanding awareness and creating new relational experiences that increase the patient's perspective. Nancy became increasingly aware of her expectations of encountering a seductive, out-of-control male and the origins of these expectations in her relationships with her father and brother. Her expanded awareness enabled her gradually to gain freedom from this restrictive perspective and enabled her to create sufficient personal space to lay claim to her own now reevaluated sexual desires.

Organizing Activity: Self and Motivational Systems

Developing Sense of Self

We begin life with innate patterns of needs and responses that coordinate the satisfaction of those needs with the responsivity of caregivers. As a consequence, a sense of self develops that becomes a center for initiating, organizing, and integrating motivation and experience (see Lichtenberg 1989, Lichtenberg et al. 1992). Affects amplify and create personal meaning in lived experience and their abstracted memories. By the third year, lived experiences in the form of events and event memories are organized as narratives. Each new perception is influenced and categorized through two simultaneous modes of processing—the logical, linguistically anchored, secondary process (generally left cerebral hemispheric functioning) and the sensory-metaphoric, imagistic, primary process (corresponding with right cerebral hemispheric processing) (Bucci 1985, Dorpat 1990, Fosshage 1983, Holt 1967, Lichtenberg 1983, McKinnon 1979, McLaughlin 1978, Noy 1979). Each lived experience is affected by the motivational system dominant at the moment and by the state of self-cohesion. Simultaneously with these intrapsychic factors, each lived experience is shaped by the pulls of an intersubjective context.

Motivational Systems

What do we mean by motivational systems? By integrating developmental research and neurophysiological studies with psychoanalytic theory and practice, Lichtenberg (1989) developed a comprehensive motivational theory. The unifying goal that he set for himself was "to define motivational systems that exist in early infancy, persist in altered forms throughout life, and characterize observable changes in motivational dominance in an analytic session" (1988, p. 60). Lichtenberg (1989) proposed five motivational systems that are built around fundamental needs, "each based on behaviors clearly observable beginning in the neonatal period: (1) the need to fulfill physiological requirements; (2) the need for attachment and affiliation; (3) the need for exploration and assertion; (4) the need to react aversively through antagonism and/or withdrawal; and (5) the need for sensual and sexual pleasure" (p. 60). These needs are "hardwired" and are present throughout the life span. They can propel us into action, or they can be aroused from the outside. For example, the neonate when alone may curiously explore a mobile or, when drowsy, be stimulated and impelled to explore the rattle that mother offers. The adult may be motivated to read and explore a book or, when drowsy, suddenly be intellectually stimulated by a colleague's telephone call and discussion. Affects play a central role in that they amplify motivations, increase the significance of the functional activity, and enhance communication.

Each motivational system begins in the neonatal period "with innate perceptual-affective action patterns and the capacity for early learned responses in coordination with inborn preferences" (Lichtenberg 1988, p. 60). The caregivers offer matching responses that foster repetition and regulation. For example, the parent who engages the baby in exploration of a new object attempts to regulate affect intensity, attention span, and interactional directionality-of-influence shifts. It is within this caregiver environment and through lived experience that the motivational systems develop and are organized. In this way hardwire givens are effectively balanced with intricate shaping of the environmental surround.

The term *motivational systems* is chosen to convey that "we are dealing not with structures or functions but with continuously ongoing processes" (Lichtenberg 1989, p. 6). While *systems* imply organization, the term emphasizes change and plasticity. Lichtenberg (1989) states that system also "conveys activity, such as organizing, initiating, and integrating, and thus fits well with the view (Wolff 1966) that infants *never* exist in a phase in which they are the passive recipient of drive pressures and environmental forces" (p. 6).

Each motivational system emerges in, is shaped by, and becomes established through lived experience within a relational context. Each system is self-organizing and self-stabilizing. Each has a regulatory effect on the other. "Positive development in one enhances motivational stability in the other" (Lichtenberg 1988, p. 61). Each system is permanent and shifts in dominance with the others from moment to moment. The functional success or failures of motivational systems affect, in turn, the sense of self. These shifts in motivational dominance have far-reaching clinical utility. The theory essentially broadens and legitimizes the range of motivations, and when using a unitary model, overcomes the tendency toward clinical and theoretical reductionism (see Fosshage 1995c).

Organizing Activity

The organizing activity we speak of does not produce replicas of prior experiences, but creates new lived experiences modeled on significant features of prior experiences. Let us consider one sequence of Nancy's lived experience in this light. We assume from the clinical evidence that Nancy's mother on her return to the home (after a period of illness following Nancy's birth) did not create for Nancy a rich, warm sense of intimacy and attachment. This experience of being cared for by a depressed? apathetic? withdrawn? mother doing her moral duty was abstracted from innumerable interactions and generalized to an expectation that Nancy could not count on a warm reception, especially from a maternal figure. By the time she was 3, these abstracted lived experiences coalesced in the narrative of Nancy pulling on her mother's skirt only to feel her mother stiffen in an attitude of refusal and rejection. When we turn to the clinical exchanges, we find that the lived experiences and the narrative "memory" shape the new creations. Nancy became aware, for example, that out of her anticipation of rejection by men, she protectively would partially withdraw and communicate unavailability.

Two Broad Groups of Self-Experiences

Experiences that contribute to vitality are particularly influential organizers of new creations, for example, the experience of the caretaker's responsive attunement to the child's need for affirmation—occurrences we refer to as *selfobject experiences* (Kohut 1977, 1984, Lichtenberg 1991, Lichtenberg et al. 1992). Selfobject experiences facilitate the psychological development of the child and motivate the adult to re-create comparable experiences of positive affect-laden attachments between self and other. In her early years, Nancy's company was welcomed by her grandfather, who would chase her brother away when he tormented her. In the analysis, Nancy sought to re-create a similar feeling of welcome and protection from the analyst.

In contrast, those experiences that involve thematic misattunements will tend to organize expectations that lead to the dominance of one motivational system, often aversiveness, at the expense of other motivations, such as attachments. For example, Nancy felt that her mother never developed a warm attachment to her, as she had with Nancy's brother, and instead, resented Nancy as a burden. Nancy's experience illustrates that an absent or denigrating parent, by disrupting the child's unique developmental process, contributes to the formation of expectations of misattunement and organizations of self and self-with-other based on aversiveness that involves affects of shame, guilt, fear, and anger. Nancy viewed herself as a person with "wrong" wishes and desires and as the responsible seducer who therefore had to constrict herself. A conflicted, debilitating identity such as Nancy's persists as an organizer because it produces a temporary sense of cohesion derived from strong affects and a sense of stability derived from a familiar sense of self. Nancy's feeling that she was in the "wrong" persisted, despite her desire to feel more positive about herself. The variety of self schemas, together with shifting motivations and interactions with others, contributes to the moment-to-moment self states and to the more pervasive and continuous sense of self.

Transference/Countertransference: Experiences of the Other Constructed by Patient and Analyst

While analysis requires placement of the patient's immediate experience in the foreground of focus, both patient and analyst enter the psychoanalytic arena with their respective prior lived experiences, shifting motivational priorities and self states, and create a unique experience with one another. The patient uses the self-enhancing and self-debilitating expectations drawn from past experience to interactionally and perceptually construct the analytic relationship. Self-enhancing patterns involve the patient's expectations that needs will be met and self-cohesion furthered (Kohut 1977, 1984, Stolorow and Lachmann 1984–1985). When entering psychoanalytic treatment, the patient's hope for a different and growth-producing experience is an attempt to self-right. Nancy described at the beginning of treatment that she tended to drift along and hoped that someone would light a fire under her. Self psychology's contribution has been to monitor closely success and failure of the patient's search for selfobject experiences and vicissitudes in the development and maintenance of a resilient, vitalized sense of self.

Analytic exploration of self-debilitating patterns and their origins generates a new perspective and gradually leads to symbolic reorganization when aided by the contrasting positive, often new, relational experience occurring in the analytic relationship. Sexual feelings for Nancy consistently triggered feelings of being wrong and the responsible seducer. As she gradually became aware of this pattern and its origins, she was able to slowly gain freedom from its dominance and to establish new, more enhancing views of herself that enabled her to become orgasmic. In other words, analysis focuses on exploring and gradually transforming the pathological configurations, while forming and bolstering the vitalizing configurations.

In contrast to viewing countertransference responses as limited to the analyst's personal pathology, we view the analyst's countertransference to embrace the full spectrum of the analyst's experience of the patient (Fosshage 1995b). The analyst's experience of the patient is constructed by his perceptions of the patient and his experience of the dual aspects of the patient's transference as well as by his prior lived experiences, motivational priorities, self states, selfobject needs (Bacal and Thomson 1996), and psychoanalytic models. Pulled both by the patient's problematic relational constructions and by the hoped-for, vitalizing wishes, the analyst's experience of the patient serves as a central guide for exploring the patient's experience.

Throughout Nancy's analysis, the analyst had to struggle with a variety of responses to her reactions to threatening or abusive situations. On one occasion, Nancy's wealthy aunt promised her a gift of a modest sum that would help Nancy not to feel pressed to overwork, and the analyst shared Nancy's relief at this prospect. He also felt pleased when after a long interval of silence Nancy said that although she was very reluctant, she would ask her aunt about the promised gift. Nancy then reported her aunt told her she had decided not to send her the money because Nancy would only waste it on her analysis. Terribly upset, Nancy called her brother in hopes of receiving solace, only to be told that he thought Nancy was a fool to do anything to risk her eventual inheritance from their aunt.

Believing he was reflecting Nancy's feelings, the analyst suggested she was disappointed, hurt, and angry at both responses. Nancy responded that at first she was angry but quickly she realized that her aunt had a right to do what she felt best with her money and was only trying to protect her and that her brother was doing the same. At this point the analyst experienced fury that quickly moved from aunt to brother to Nancy with a generous helping of impotence thrown in. Nancy went on to ask why her aunt should give or leave her anything anyway. What claim did she have on her aunt for either money, help, or concern? At this the analyst's sense of outrage powered his entering the role enactment taking place between them with his own value judgment: "You are her sister's daughter, she is your aunt."

The analyst's reaction can be seen as an amalgam of identification with Nancy's suppressed rage, as a direct attack on her for her passivity in response to being teased with unkept promises, for turning to her brother despite the knowledge that he would usually take any opportunity to humiliate her further, and for the slur on the worth of analysis that constituted the family attitudes—a personal reason of the analyst to remind Nancy of family values. We do not believe the outcome of this countertransference experience of the patient is predictable a priori as either destructive or helpful. The sequence that followed indicated that Nancy experienced it as an empathic confirmation of her having a place in her family that called for responses of sensitivity and concern.

Selfobject Experiences within the Analytic Relationship

The analysand's hopes and expectations of evoking selfobject experiences from the analyst are fundamental motivations for the analysand's engagement in the analytic

endeavor. Self psychology has emphasized mirroring, twinship, and idealizing selfobject transferences, all of which are
cornerstones of the attachment motivational system. We
believe that the application of the motivational systems theory enlarges the types of selfobject experiences that emerge
within the analytic relationship.

Regulation of Physiological Requirements

Self psychologists have demonstrated that disturbed regulation of physiological requirements in patients can result
from primary disturbances or deficiencies in mirroring, twinship, or idealizing experiences. This conception was based on
the observation that, when a sense of empathic connection
was restored after a disruption, the physiological disturbance
would end. Nancy frequently would report periods of consti-
pation over weekends that would abate when a mirroring,
twinship, or idealizing transference experience was restored.
In contrast, many disturbances of eating, eliminating, sleep,
breathing, and equilibrium appear to be primary defects in
lived experience. Some primary disturbances are the result of
innate dysregulation such as infantile asthma, eczema, tactile
sensitivity, sensory hyperreactivity, and so on. More com-
monly, developmental disturbances are the result of failures
in coordination between caregivers and child. Once specific
failures in physiological regulation occur, they frequently
become the source of disturbances in attachment experiences.

At times the interweaving of primary dysregulation and
attachment deficits are difficult to disentangle. Nancy's infan-

tile eczema is probably a primary skin disturbance but was associatively linked to her sense of failure to be held and fondled sensually by her mother. Whenever her eczema recurred during the analysis, she experienced discomfort, embarrassment, and an intense craving for being held physically by the analyst. The analyst could not discern if the craving that had been unresponded to triggered the eczema or the eczema triggered the craving. When there is a long-standing dysregulation that is current (for example, a dysregulation of hunger and satiety or sleep disruption), the affect states related to these disruptions, and the current motives related to them, require specific empathic focus to understand the particular failure of selfobject experiences suffered by the patient. This understanding will increase perspective and aid the patient in achieving physiological regulation in the affected areas.

Attachment and Affiliation

In addition to affirming, twinship, and idealizing attachment, selfobject experiences can include a variety of features that can be more specifically referred to as guide, advocate, mentor, sponsor, lover, and rival. Nancy, for example, looked to her analyst as a mentor to help her to regulate herself in a variety of ways, including her relationships to men, her graduate work, and her finances. Affiliative selfobject experiences can involve the family, team, country, religion, and professional group—all with specific allegiances to values and ideologies. Religious affiliation was important to Nancy as a positive source of intimacy. Rather than her nuclear and extended family, where she felt criticized and rejected, she looked to religious groups for inclusion and acceptance. This led to her search within the intersubjective atmosphere of the analysis for her reading of the analyst's position. Was he opposed to her conversion to Catholism, as was her family? Was he opposed to religion as a "neurosis" as she knew some psychoanalysts were? She knew that he did not work on Rosh Hashanah and Yom Kippur and so inferred he would be supportive. She concluded from the way that he responded to issues she brought up that he wished to help her achieve a positive sense of affiliation.

Exploration and Assertion

Exploratory and assertively motivated behavior can create a pleasure of efficacy and competence, an inherent self-enhancing aim. Successful exploration and assertion, exercising talents and skills, create a selfobject experience. Past experiences accruing to this motivational system will trigger attitudes that will affect the patient's willingness to articulate and explore their internal process in the analysis. When Nancy entered treatment, she had begun doctoral studies and was well aware of the vitalizing experience of using her talents. She spent many sessions discussing problems involving learning, teaching, writing, and test taking. Based on her early strong exploratory interest and the encouragement she received from her father to inquire, she easily established an investigative approach with the analyst. She enjoyed the sense of efficacy that she gained from understanding meanings and solving problems in and out of the analysis. As with her father and brothers, her faith in her ability to do exploratory work was easily lost if a professor or the analyst failed to offer her affirmations or if she discovered or inferred a prejudice against women. Then she might resort to the use of her intelligence she had employed in adolescence to set the man up to make opinionated statements she could ridicule. With the analyst, she both admired his carefully thought-out interventions and gently derided his slowness as diffidence. Alternatively, if she felt he was ahead of her in drawing a conclusion, she often experienced a strong sense of shame, of being exposed as not quick like her brother. As she became less plagued by negative self-feelings, particularly feeling "in the wrong," and by her desire to rush and dazzle, she was better able to pursue her work with clearer focus and to navigate relationships concerning her dissertation with more effective assertiveness.

Aversive Antagonism or Withdrawal

To be able to become angry and to avert a perceived threat or hurt can both protect the individual and self-cohesion as well as enhance feelings of power. In augmenting exploration and assertiveness in overcoming obstacles, aggression can increase our sense of efficacy. The child needs an empathic parent to be both ally and adversary (Lachmann 1986, Wolf 1980). As allies, parents confirm that, through their children's vigorous statements of preference and opposition, their children strengthen their developing sense of self. As adversaries, parents provide children with a firm, indestructible opponent against whom to mobilize forces of anger, reasoning, and persuasion. The combination of ally and adversary gives children the opportunity to learn the sense of power that derives from augmenting assertion with anger, and thus to be effective in controversy. Similar experiences within the analysis can vitalize the patient's sense of self.

Nancy frequently would return home from a long, exhausting weekend at the laboratory and begin to experience panic that she had made an error in a test that would be fatal to a patient. She would have to call and be reassured. From the aversive transference position with the analyst that he asked too much of her while his help was inadequate, Nancy and he drew the inference that she also felt this way at work. Detailed inquiry into her relationships at the job in the laboratory that she had first described as superior indicated she had allowed co-workers and supervisors to overload her with the more difficult procedures. Her great sense of responsibility for the patients and her high standards of efficiency made her vulnerable to the flattery of the supervisor's importunings. Behind that lay her resentment that she had had to take care of her mother during migraine episodes often triggered by her mother's critical reactions to her. Added to this dually aversive experience in her early life with mother was her sense of guilt at not looking after her mother in her final illness. As these multiple threads of aversiveness were being explored, she began to protest directly to her co-workers and supervisors and institute a policy of recognizing and caring for her own needs. The controversies that followed sometimes distressed her, but her panics disappeared.

Sensual Enjoyment and Sexual Excitement

Motivations can center on seeking sensual enjoyment or on sexual excitement. Sensual enjoyment can diminish the intensity of overstimulation or aversiveness by soothing, calming, preparing for sleep, or it can increase receptiveness to erotically arousing stimuli, leading to genital arousal and sexual excitement. A patient might seek sensual enjoyment in the analytic relationship and become especially attuned to gently rhythmic vocal tones, the restful ambience, the aesthetically pleasing decor, and symbolically feeling touched and comforted. In her search for the sensual that was so absent in her relationship to her mother, Nancy, for example, would either look at or fantasize about the round, warm breasts and bottoms of women. Gradually her sensuality could also lead to sexual arousal.

Traditional psychoanalytic theory has taken sexual excitement and orgastic discharge as the central, even sole, aim of the sensual-sexual motivational system. However, we have noted that when a person feels endangered, sexual excitement may be sought, not as a primary goal, but as a means to obtain vitalization to repair a depleted state. Although this often might be the case, sexual excitement as it emerges within the analytic relationship can clearly offer self-enhancing pleasure. Nancy gradually and tentatively allowed herself to experience sensual longings and sexual arousal toward the analyst, both away from and during the sessions. These experiences carried their own weight in the treatment as the working out of problems in the sensual-sexual motivational system, rather than being solely or primarily displaced attempts to desperately repair attachments. The repair occurred directly in the recovery of sensation in her genitals after years of anesthesia. Following this important gain in sensuality, she had her first experience with orgastic excitement. The overall result was that for the first time, she felt like a "complete woman."

Exploring the Transference Experience within an Intersubjective Context

We have stated earlier that the patient and analyst variably co-determine the creation of a transference experience and that the range of contribution for each varies from minimal to considerable. We now address two possible errors that can be alleviated by the analyst's awareness of the variable range of contributions of both participants. In one direction, we may err when we ascribe the patient's current experience as exclusively the product of the patient when the analyst has contributed significantly. In the other direction, we may err when we insistently seek for significant origins of the patient's experience in the analyst's responses or attitudes when the analyst has contributed minimally.

When Nancy's analyst stumbled into a role enactment and pressured her to address additional meanings of a dream, interactionally creating seducer–seduced roles, he reported initially ingenuously denying his responsibility for the pressure. Nancy perceived (did not project) his ingenuousness and became angry at him for having denied his responsibility.

If he viewed her percept as a projection, he would again deny responsibility, thus continuing to replicate a pathogenic scenario. His recognition of his contribution to the interaction advantageously positioned him to validate her perception. Through subsequent acceptance of her anger and understanding and acknowledging his contribution to it, he was able to facilitate the repair of the rupture. Alternatively when patients are preoccupied with troubling expectations they regard as contrary to what their experiences with the analyst's responses have been, the old conflict-laden and the emergent views are maintained simultaneously. Nancy struggled between percepts of her analyst as either a trustworthy or an "awful" human being. While the analyst's responses, over time, reinforced her hoped-for experience and emergent percept of the analyst as trustworthy, the patient's older expectation at this moment was not contributed to by the analyst, but was primarily generated intrapsychically. An analyst's assumption and insistence on his participation would, in this instance, obfuscate the patient's intrapsychic struggle with these two percepts.

The process of psychological reorganization breeds cognitive dissonance (Festinger 1964). The struggle with cognitive dissonance is amply demonstrated in Nancy's questioning: "I'm not at all certain you're not just an awful human being. Who are you? Am I doing the right thing by coming? Can I trust you? Yet, I find myself not wanting to be separated from you." Interpretations that are too dissonant with the patient's percepts will meet with aversiveness and disrupt analytic exploration. Remaining close to a patient's experience and gradually introducing a new frame (interpretation) will enable a patient to remain open to and gradually assimilate a new perspective (understanding).

The complex manner in which a patient organizes his or her experience of the analyst within the analytic relationship usually can be illuminated only after repeated efforts of recognition, conceptualization, and revision. Through empathic inquiry, patient and analyst over time identify a repetitive experience occurring within the analytic relationship. The patient's description of the same experience in other relationships further validates that it is thematic. Often we can begin to explore the history of the experience, and, on occasion, a patient easily leads us to seminal experiences (model scenes) of the past. In other instances, the patient's experience of the analyst is so intense that the experience can only be "lived in" by analyst and patient (what we refer to as *wearing the attributions*). To wear the attributions means to explore and to fill in as if the attributions are true (to the patient they are true). Any exploration that moves away from the ongoing intense experience in the analytic relationship, whether it be identifying a theme or historical antecedents, can be experienced as invalidating the patient's current perceptions. If patients feel a loss of the validity of their experience, they will tend to avert further exploration. In contrast, analysts with a nondefensive willingness to wear the attributions, as well as timely acknowledgment of their own contributions, enable the analysand to feel heard and, in turn, to become more reflective. With the analysand's perceptions validated (that is, identified as his or hers and having a "reality"), and with increased reflective space, empathic inquiry toward identifying the experience as repetitive and exploration of the analysand's contribution to that experience can gradually proceed.

Expectations derived from past traumatic experiences may be triggered in the analytic relationship when the analyst's repeated interactions confirm those feared expectations. The consistently intense aversive experiences that follow may jeopardize the conditions necessary for the analysis of an unfolding pattern. The sense of a background selfobject experience may be disrupted, and the intensity of the affect state may render cognitive processing and reflection impossible. Analysts who can recognize that a background selfobject experience has been disrupted can be alert to modifying their behavior in order to decrease their contribution to the transference. For example, a patient's proneness to feel intruded upon and obliterated, eliciting intense aversive reactions of withdrawal, will require the analyst to become less verbally active so that the work can proceed and the patient's proneness to feel intruded upon can be explored. Kohut (1977) notes that for certain patients the understanding phase of the analysis needed to be extended before the explanatory phase in order to create the requisite developmental experiences and, we add, to offset pathogenic experiences. We believe that for all successful analysand–analyst pairs, some modifications in the analyst's behavior occur, often unconsciously, as part of their ongoing mutual regulation (Beebe and Lachmann 1992, Fosshage 1995a, Jacobs 1991). In working with patients where a seriously injurious experience is being re-created in the treatment, understanding the patient's plight may need to be conveyed through other actions, where words alone will not suffice (Bacal 1985, Balint 1968, Fosshage 1995a, Jacobs 1991, Lichtenberg et al. 1992, Lindon 1994, Malin 1993).

Conclusion

In conclusion, transference, as we define it, refers to the analysand's experiences of the analytic relationship and the organizing patterns through which they are constructed and assimilated. While analysis requires placement of the patient's immediate experience in the foreground of focus, both patient and analyst enter the psychoanalytic arena with their respective prior lived experiences, shifting motivational priorities, and self states, and create a unique experience with one another. The patient variably constructs and assimilates the analytic relationship into repetitive self-enhancing and self-debilitating expectations that were established through past experience. The analysand's hopes of evoking the selfobject experiences necessary for ensuring cohesion and vitality of the sense of self serves as the fundamental motivation underlying the analytic endeavor. Applying motivational systems theory to enlarge and specify the types of selfobject experiences sought within the analytic relationship, we believe, positions analysts advantageously to understand the shifting motivational priorities of their analysands.

References

Bacal, H. (1985). Optimal responsiveness and the therapeutic process. In: *Progress in Self Psychology*, vol. 1, ed. A. Goldberg, pp. 202–227. Hillsdale, NJ: Analytic Press.

Bacal, H., and Thomson, P. (1996). The psychoanalyst's selfobject needs and the effect of their frustration on the treatment: a new view of countertransference. In *Basic Ideas Reconsidered, Progress in Self Psychology*, vol. 12, ed. A. Goldberg, pp. 17–35. Hillsdale, NJ: Analytic Press.

Balint, M. (1968). *The Basic Fault*. London: Tavistock.

Beebe, B., and Lachmann, F. (1992). The contribution of mother–infant mutual influence to the origins of self- and object representations. In *Relational Perspectives in Psychoanalysis*, ed. N. Skolnick and S. Warshaw, pp. 83–117. Hillsdale, NJ: Analytic Press.

Bucci, W. (1985). Dual coding: a cognitive model in psychoanalytic research. *Journal of the American Psychoanalytic Association* 33:571–607.

Dorpat, T. (1990). The primary process revisited. *Bulletin of the Society of Psychoanalytic Psychotherapy* 5:5–22.

Festinger, L. (1964). *Conflict, Decision, and Dissonance*. Stanford, CA: Stanford University Press.

Fosshage, J. (1983). The psychological function of dreams: a revised psychoanalytic perspective. *Psychoanalysis and Contemporary Thought* 6:641–669.

——— (1994). Toward reconceptualizing transference: theoretical and clinical considerations. *International Journal of Psycho-Analysis* 75(2):265–280.

——— (1995a). Interaction in psychoanalysis: a broadening horizon. *Psychoanalytic Dialogues* 5(3):459–478.

——— (1995b). Countertransference as the analyst's experience of the analysand: the influence of listening perspectives. *Psychoanalytic Psychology* 12(3):375–391.

——— (1995c). An expansion of motivational theory: Lichtenberg's motivational systems model. *Psychoanalytic Inquiry* 15(4):421–436.

Gill, M. (1982). *Analysis of Transference 1: Theory and Technique.* New York: International Universities Press.

——— (1983). The interpersonal paradigm and the degree of the therapist's involvement. *Contemporary Psychoanalysis* 19:200–237.

——— (1994). Transference: A change in conception or only in emphasis? *Psychoanalytic Inquiry* 4(3):489–523.

Hoffman, I. Z. (1983). The patient as interpreter of the analyst's experience. *Contemporary Psychoanalysis* 19:389–422.

——— (1991). Discussion: toward a social-constructivist view of the psychoanalytic situation. *Psychoanalytic Dialogues* 1:74–105.

——— (1992). Some practical implications of a social constructivist view of the analytic situation. *Psychoanalytic Dialogues* 1:74–105.

Hoffman, I. Z., and Gill, M. (1988). Critical reflections on a coding scheme. *International Journal of Psycho-Analysis* 69:55–64.

Holt, R. (1967). The development of primary process. In *Motives and Thought: Psychoanalytic Essays in Honor of David Rapaport*, pp. 344–383. (Psychological Issues, Monograph 18/19). New York: International Universities Press.

Jacobs, T. (1991). *The Use of the Self.* New York: International Universities Press.

Kohut, H. (1977). *The Restoration of the Self.* New York: International Universities Press.

———— (1984). *How Does Analysis Cure?* Chicago: University of Chicago Press.

Lachmann, F. (1986). Interpretation of psychic conflict and adversarial relationships: a self-psychoanalytic perspective. *Psychoanalytic Psychology* 3:341–355.

Lachmann, F., and Beebe, B. (1992). Representational and selfobject transferences: a developmental perspective. In *New Therapeutic Visions, Progress in Self Psychology,* vol. 8, ed. A. Goldberg, pp. 3–15. Hillsdale, NJ: Analytic Press.

Lichtenberg, J. (1983). *Psychoanalysis and Infant Research.* Hillsdale, NJ: Analytic Press.

———— (1988). A theory of motivational-functional systems as psychic structures. *Journal of the American Psychoanalytic Association* 36, Supplement:57–72.

———— (1989). *Psychoanalysis and Motivation.* Hillsdale, NJ: Analytic Press.

———— (1990). Rethinking the scope of the patient's transference and the therapist's counterresponsiveness. In *The Realities of Transference: Progress in Self Psychology,* vol. 6, ed. A. Goldberg, pp. 23–33. Hillsdale, NJ: Analytic Press.

———— (1991). What is a selfobject? *Psychoanalytic Dialogues* 1(4):455–479.

Lichtenberg, J., Lachmann, F., and Fosshage, J. (1992). *Self and Motivational Systems: Toward a Theory of Technique.* Hillsdale, NJ: Analytic Press.

———— (1996). *The Clinical Exchange: Techniques Derived from Self and Motivational Systems.* Hillsdale, NJ: Analytic Press.

Lindon, J. (1994). Gratification and provision in psychoanalysis: Should we get rid of "the rule of abstinence?" *Psychoanalytic Dialogues* 4:549–582.

Malin, A. (1993). A self psychological approach to the analysis of resistance: a case report. *International Journal of Psycho-Analysis* 74:505–518.

McKinnon, J. (1979). Two semantic forms: neuropsychological and psychoanalytic descriptions. *Psychoanalytic Contemporary Thought* 2:25–76.

McLaughlin, J. (1978). Primary and secondary process in the context of cerebral hemispheric specialization. *Psychoanalytic Quarterly* 47:237–266.

Noy, P. (1979). The psychoanalytic theory of cognitive development. *Psychoanalytic Study of the Child* 34:169–216. New Haven, CT: Yale University Press.

Ornstein, A. (1974). The dread to repeat and the new beginning. *The Annual of Psychoanalysis* 2:231–248. Madison, CT: International Universities Press.

Stern, D. (1985). *The Interpersonal World of the Infant.* New York: Basic Books.

Stolorow, R., and Lachmann, F. (1984–1985). Transference: the future of an illusion. *The Annual of Psychoanalysis* 12(13):19–37.

Stolorow, R., Brandchaft, B., and Atwood, G. (1987). *Psychoanalytic Treatment: An Intersubjective Approach.* Hillsdale, NJ: Analytic Press.

Wachtel, P. F. (1980). Transference, schema and assimilation: the relevance of Piaget to the psychoanalytic theory of transference. *The Annual of Psychoanalysis* 8:59–76.

Wolf, E. (1980). On the developmental line of selfobject relations. In *Advances in Self Psychology*, ed. A. Goldberg, pp. 117–132. New York: International Universities Press.

Wolff, P. (1966). The causes, controls, and organization of behavior in the neonate. *Psychological Issues*, Monogr. 17. New York: International Universities Press.

5

A Kleinian View of Early Development as Seen in Psychoanalysis in Children and Adults

Patricia Daniel

In the tradition of psychoanalysis starting with Freud, we develop general hypotheses about early mental life from our analysis of patients, adults, and children. Melanie Klein studied the mind of the infant through the analysis of small children, and she became increasingly aware of the complexity of the processes operating in the early stages of development. So impressed was she by her experiences with her young patients that by 1936 she wrote "Infantile feelings and phantasies leave, as it were, their imprints on the mind, imprints which do not fade away but are stored up, remain active, and exert continuous and powerful influence on the emotional and intellectual life of the individual" (p. 290).

While we do not *know* what goes on in the mind of the baby, Kleinians hypothesize that primitive phantasies develop from birth onwards and that a rudimentary ego starts to form through contact with characteristics of the parents, such as smell, tones of voice, ways of holding, and other features of the manner in which the mother cares for the baby. The infant brings its own unique genetic endowment, its physical constitution, and its instinctual life. During pregnancy mother has conscious and unconscious phantasies about her baby, and their postnatal experiences together will tend to confirm or modify her expectations. We would all agree that the early mother–child relationship lays the foundation for all subsequent relationships. I shall return to this, but first I wish to consider a brief extract from an observation of a mother and baby, to focus attention on the intricacy of the detail of that interaction, and to emphasise how observation and imagination stimulate ideas and thoughts about what and how the baby may be experiencing and feeling. The observations took place weekly in the home of the mother and baby.

A Feeding Baby Observed

A significant feature of the early months in the life of baby Sally was her mother's close attunement with this baby, her first. It seemed that mother responded most sensitively to the baby's moods, but when the baby was distressed and appeared to be in a relatively unintegrated or disintegrated state, mother's anxiety greatly increased. Sally was 9 days old at the time of this account.

> Mother sits in what appears to be an uncomfortable position. (I wondered whether this was an indication of her feelings about looking after the baby and of being observed.)
>
> She held the baby awkwardly in her left arm while uncovering her left breast and preparing the nipple: Sally's mouth was wide open, and she was also trying to push her mouth forward. When her face reached the breast, her nose became buried in it and her lower jaw seemed to be trying to open wider as if to take in the entire breast. Mother drew back, tried to get the nipple to stand out and express milk, and she again brought the baby's head to it. Again the same thing happened. Sally's right hand, which was enveloped in two inches of a much too big sleeve, waved round the upper part of the breast but kept coming back to the point where the baby's mouth was against the breast. Several attempts followed, and it was difficult to see what was going on because of the tangle of baby clothes, breast, and Sally's face and mouth. Mother was becoming agitated and said that this was

the difficult side (implying the left breast) but, she said, it had never been as bad as this. By now the baby was moving her head from side to side in a frantic attempt to find what she wanted, and her face was bright red. But still she did not cry. Eventually, instead of having it wide open, she started to pout her mouth and then it narrowed sufficiently for her to take hold of the nipple; she started to suck slowly and rhythmically. Her legs and feet became still, and her right hand wandered over the breast.

Later, after some burping, mother put the baby to the right breast and the report continues:

Sally got the nipple into her mouth at once and for a minute or two held it in her mouth without sucking. Then she sucked slowly and steadily: after a while she stopped again, still with the nipple in her mouth, and then continued sucking. It was striking how mother and baby now seem to fit into each other. Sally was held in a good position in relation to the breast; her legs were relaxed and her right arm rested on the breast near the nipple and her mouth. Mother, too, looked comfortable as she talked to me about the earlier feeds that day, how she feels tired but how good Sally is. She looked at the baby, murmured endearments, and then resumed talking to me about how the midwife and Health Visitor were helpful, not a bit bossy. Suddenly mother jerked and exclaimed that Sally bit her. The baby immediately froze into rigid stillness, including her mouth. . . .

This glimpse of their relationship is, from their point of view, already quite late on—the ninth day. There have been many feeds and mother knows feeding from the left breast is more difficult than the right. Could this already be the beginning of a differentiation in the baby's mind between a frustrating "bad" breast when she wants it but it is not there and a satisfying "good" breast as she begins to suck? When the baby bites, might mother in the course of talking have withdrawn the nipple a little and so Sally clamped down? When Sally does bite, mother's sharp jerk and sudden exclamation are met with a counterresponse in which the baby freezes. May such a reaction, over time, develop into a characteristic initial response to stress? (I shall return to this later.)

Then there is mother's anxiety over the difficult left breast and her subsequent remarks about the midwife and Health Visitor. These remarks, this negation, suggest an underlying fear that they might be bossy. While these may, in part, be based on realistic expectations of midwives, they may also be transferred from mother's early experiences, roles and expectations, and fears of being bossed. In fact, she is reassured by the reality of the experience with the midwife and Health Visitor. By the same token she may transfer such expectations onto the observer, feel reassured by her, and then feel reassured in her handling of the baby. I have mentioned earlier mother's phantasied expectations being confirmed or modified by her actual experiences. Note too that this mother already talks about Sally as a person with her own little ways, in contrast to mothers who tend to see their babies as extensions of themselves. So she is not only dominated by her expectations but is also able to meet the "real" baby, midwife, observer. Thus we can use imaginative conjecture about the nature of the baby's early experience, even though we cannot know objectively the baby's subjective experience. We also have to guard against projecting our expectations, then to be confirmed by what we see. But there is, as it were, space in the mind of the observer to monitor these and be imaginatively alive to what might also be the nature of the baby's subjective experience, bearing in mind that there is in all science no pure observation.

Observational studies of infants (Cohn and Tronick 1983, Murray and Trevarthen 1985, Stern 1985, 1991, 1993) provide us with increasing knowledge about the infant's physical and mental capabilities. They raise questions as to what and how much is given in innate structures, the constitutional element, and therefore how far infants are pre-adapted to the external world, as against how much is constructed from birth onwards—the environmental factor. I am interested particularly in the interaction of these forces. These studies contribute to developmental perspectives and alert us to developmental divergence. That 2-day-old babies can imitate the facial expressions of others (Field et al. 1982) and 2- and 3-month-old babies avert their eyes and bodies and show distress when their mother's faces go deadpan, should, I think, encourage us to speculate about the content of their minds. When we try to do so, we come up against some limitations: the constraint of language to convey what we think goes on in the infant's mind and the difficulty in making comprehensible to ourselves what the infant experiences but does not understand. We are all astonished by what we observe, and yet there is also a mystery about what goes on inside the baby. Analysis is the only tool we have to find out about the mystery of the infant's inner life, and we do believe, from our analysis of children and adults, that infants have an internal life. It cannot be deduced from observational studies.

As the developmentalists have come to think, what Klein inferred from her analytic work is that an emotional relationship between mother and infant comes into being from birth onward. She assumed that the infant has the capacity to relate to an object from the start and that this presupposes a rudimentary ego capable of experiencing anxiety and able to react defensively against it. In this view, the infant directs its impulses, feelings, and reactions toward an object, initially its mother. Thus the development of the ego begins within the context of the primary object relationship. Klein follows Freud in postulating two inborn drives, the life and death instincts that underpin all mental life. Their derivatives, love and hate, are thought of as tending toward integration and/ or disintegration. Kleinians assume that the infant, in part, experiences its drives as "good" or "bad," and attributes these to its objects—and the "bad" arouse anxiety. Segal (1964) puts it graphically when she says the baby "tastes" his experiences and classifies them as either good or bad.

Returning to baby Sally, some might say the baby biting, then freezing in response to mother's jerk, are purely mechanical reflex reactions. Fraiberg (1987) describes *freezing* as a defense against danger. I think Kleinians would say it contains or includes primitive mental reactions, the biting being an action of the baby, and the freezing a defense against anxiety caused by mother's jerk, and that these have some phantasy content, however primitive, of what the baby feels she does to the object and what she feels the object then does to her. For example, we might speculate that the baby's perception of mother's jerk was retaliation to the baby's bite. As soon as we put this into words, give it language, it may make it more than what it is, but just to think of it as reflex, I believe, may make it less than what it is. Anxiety has to be gotten rid of and is expelled, in phantasy, back into the object by projection. The first "bad" experience has long been thought to be the trauma of birth, although with the expanding interest in neonatal life (Piontelli 1992), there are growing speculations that experiences in utero may already have made their mark on the infant psyche.

For the infant, the first weeks of life are, in part, a struggle for physical and psychic survival, integration versus disintegration, love versus hate. This struggle, we believe, takes place within relationships with objects. During the first three months or so, the parents are in part experienced by the infant as many disconnected aspects and functions—part objects—and this is due to limitations in perception and to the need to keep good and bad apart by splitting the object. When the object is split, the ego splits too—one part in relationship with the good object and the other part in relationship with the bad object. This early splitting is of importance for normal development because it provides a primitive structure to the ego and allows the infant to organize his chaotic experience and manage his anxieties. When the infant is repeatedly exposed to excessive anxiety, pathological forms of splitting may develop (Bion 1957, Klein 1952b, Rosenfeld

1950). The early split between good and bad aspects of the infant self and its objects will form the basis for subsequent loving and hostile relationships.

There is also evidence of the integrating function of the ego—studies, for example, which show that infants recognize their mother's breast pad at 6 days old (MacFarlane 1975) and the integrating function of the bad object. Does baby Sally have a "memory" of the "bad" left breast and recognize it in some way? If the baby can recognize or remember, is there not some mental image in her mind, and might this indicate the existence of early phantasy (Brenman Pick 1992)? During the first three months we think the baby's mind functions predominantly in the paranoid-schizoid position, moving between states of idealized union with its objects, experienced as part objects, or feeling terribly persecuted by them. The ego is immature and relatively unintegrated, but making moves toward integration.

Bion (1962) offers the hypothesis that what he calls the mother's *reverie* forms the psychological fountain of the supply of the infant's need for love and understanding. He suggests that if, for whatever reason, reverie is not present or is not associated with love of the infant and its father, then this psychological fact will somehow be communicated to the infant, even though this may be incomprehensible to him. Bion, following Klein, assumes that all the mother's nurturing links with the infant serve as channels of psychical communication between the two—a psychosomatic experience. In Bion's terms, the mother's reverie is a mental state in which she is able to receive the baby's projections, whether felt as good or bad by the infant. When she herself is distressed or agitated, her psychic balance and reverie are disturbed so that she then becomes unable to introject, to unconsciously process into understanding, and respond to the baby in a way that is emotionally helpful. Should this state of affairs be, or become, a pattern, for example in states of depression or mania, the baby is driven to project its experiences more forcefully, even violently, or to withdraw in despair. Either response over periods of time may lead to disturbance in the infant's subsequent object relations.

Through minute and repeated projection and introjection of experiences felt as either good or bad, an internal world builds up of self in relationship with internal and external objects, and objects in relationship with each other. In the Kleinian view this dynamic internal world is subjectively experienced in the form of unconscious phantasies: we think the very earliest phantasies are probably experienced as bodily sensations and that only gradually physical and mental become differentiated.

Melanie Klein (1935, 1940, 1946) made an important contribution with her formulation of two basic constellations of phantasies and relationships to objects, together with their characteristic anxieties and defenses: the paranoid-schizoid and the depressive positions. While developmentally the paranoid-schizoid position is the more primitive and predominates in the first months of life and the depressive position comes more to the fore around the fourth month, the two positions coexist, even in the earliest months. They continue throughout life and can fluctuate in their dominance in the mind at any one time. They are not seen as stages of development to be passed through or outgrown. In the paranoid-schizoid position the individual alternates between states of idealized union with objects felt to be excessively good and feeling persecuted by objects felt to be excessively bad.

The dominant anxiety is for the survival of the self and the characteristic defenses of the position are splitting, projective identification, and idealization. Objects are perceived and experienced as part objects, partly due to limitations in perception and partly because the ego is relatively unintegrated and splits its objects. The infant is, from the very beginning, threatened by destructive impulses from within, and to defend its loving self, it projects the hostile forces into its object, whom it then fears and feels hates it. For example, when the baby bites, in phantasy it projects its aggression into the breast. When mother flinches, the baby then feels it is the breast/mother who bites *it* in retaliation, so the baby freezes out of fear. Klein (1952a) described the infantile neurosis "as a combination of processes by which anxieties of a psychotic nature are bound, worked through and modified" (p. 81). That is to say the anxieties experienced by infants in the normal course of development are of the intensity and quality that we see in psychotic states in adults. This is not to say that Kleinians see all babies as psychotic.

Clinical Material

I hope the following clinical material will show how that which remained so mysterious in the observation was made visible in a psychoanalysis. The material is from a session in a child analysis when James was approaching his ninth birthday. He had one sibling, a brother a few years older than himself. The approach to his birthday brought forward phantasies about his birth that had been seen and worked on earlier in the analysis, but they now reappeared compressed within the space of one session.

On entering, James makes two remarks: that he has a cold and is wearing a new T-shirt. This indeed looks huge on him. Taking the jug of water and basin, he seats himself at the table, drinks a little but most he pours from below his nose so it trickles over his mouth: some he catches in his mouth and the rest runs down into the basin. These actions resemble a baby perhaps both attempting to get fed and/or not wanting to feed. He continues this for about five minutes before wiping his face on the couch cover and pillow. Next he pulls the T-shirt down to cover his bottom and the sleeves to cover his hands, lies on the couch with the blanket over him, the pillow under his head, swivels round to place his feet on the table, and lies there sucking a lollipop.

I say that first he felt undecided about what he might get or want from analysis today, but now he wishes to be tucked away inside, as in his T-shirt—even the table that he rests on has become like a part of me. "No, no, I don't. You look stupid, you're a troublesome brat," he replies. But he moves the table nearer to me and moves himself entirely onto it, lying on his tummy facing me and holding the pillow to his chest. In this position he alternates between making faces at me and cuddling his face in the pillow. I say that I have become the stupid brat because he believes that if I realize that he feels cold and wants to cuddle up, I'll think him small and stupid. He teases me so that I won't notice that he feels as though he has now got onto my lap, as though the table is my lap. This time there is a different response. He blows his nose and puts another sweet into his mouth. Then he transfers himself, without touching the floor, onto a swivel arm chair and pulling the blanket round his neck says he'll make himself cozy and warm. He swivels the chair so his back is to me.

Initially he is an 8-year-old who knows he is separate and reports his cold. I believe he also feels that his analyst has left him out in the cold since yesterday and that he comes back to a "cold" analyst today. He then draws attention to his new T-shirt, I think suggesting that he feels he has to be, or wants to be, bigger as he comes closer to age 9, hence the huge T-shirt. This may be both the drive toward real growth and development and the attempt at mammoth antigrowth by projective identification—being inside the object and taking possession of it or being possessed by the object. We can speculate about a similar movement in the observed baby: her drive toward life in the sucking and feeding, and the antigrowth in the attempt to take possession of the whole breast in one gulp. We can also speculate about baby Sally's fusion or enmeshment with mother in the tangle of sleeves and mother's clothes, as James is lost in his huge T-shirt. At the start of the session James feels he returns to a "cold" analyst, and baby Sally at the start of the feeding to a breast she seems to want but that initially frustrates her. This raises the question about the nature of the object the baby feels she has returned to.

To go back to James, he starts to enact his phantasy of being an "inside" baby, but when this is interpreted, I think that he hears the analyst as a primitive, cold superego who hates him and attacks him as a boy who should not have these wishes. This would link with the analyst who he feels had left him out in the cold. So he immediately projects the baby him into the analyst who looks stupid, while he becomes the adult with a troublesome brat of a baby/analyst. All the subsequent movements without touching the floor could be understood as confirmation of the omnipotent quality of his phantasy of being inside his object, so he never has to touch ground (or reality) as a separate person. But soon the omnipotence weakens, and he becomes both the baby burying his head in the pillow/breast and the 8/9-year-old teasing his analyst. Did the interpretation to this effect increase his anxiety about separateness and make him feel *he* was being teased by the analyst in retaliation? For he goes back inside via the couch into the swivel chair with his back to me. Or is it that he turns his back on an analyst who understands him, driven by his "cold" hatred of dependence on an understanding object? If this were so, it would indicate a deep split in his infantile self. Or is the problem for the latency boy nearing age 9 that in the move toward growing up he can only bear to stay with his infantile self briefly and then he has to turn his back on it or he would never give it up? This sequence also raises questions about his history: Might there have been either a cold superego mother or an overindulgent mother who would never support him to grow?

In the following further material from later in the same session, it seems to me that a phantasy emerges about the damage done to his object during his birth. The enactment arouses such anxiety that it precipitates curiosity about his object; then an unexpected external event throws him into confusion and acute anxiety.

A minute or so later he tosses off the blanket, swings the chair round so we can see each other, and scrapes both his shoes up and down the wall, saying he has "walked in red paint and look what's happening to the wall." After much swiveling and scraping of shoes on the wall, I hear snatches of broken sentences followed by several questions addressed to me about myself. Then suddenly we hear children outside calling, "Come inside, come on in." James looks terrified and then, confused, mutters "Children out there?" and takes a hanky from his pocket, smells it, says it's his brother's and blows his nose with it. Once more he transfers himself back onto the couch, without touching the floor, as though again going back "inside." We are reminded of baby Sally's "freeze" by James's panic: both may indicate early panic/confusion as to where the attack is coming from.

When he scrapes the wall with his shoes, mentions walking in red paint, and directs the analyst's attention to the immediate damage he is doing to the wall, is he wanting to make the place unfit for other children? I think the calls from the children outside panicked him. Were they friendly or specters of the babies he believes he kept out in this way? Then another phantasy is stirred about what he did to his mother's insides. The intrusion of the children from the outside into the session, via their calls, also stimulates jealousy and fear of being replaced and may drive him toward vengefulness. But he may also be projecting onto the children his own impulses to attack mother from inside and outside, and these wishes, in the immediacy of the transference, become the threat to the analyst whose walls are being bloodied, and whose interpretations are being broken up in his snatches of broken sentences.

In such a clinical situation we might wonder how the father is experienced. In actuality his mother maintained an exceedingly close relation with James while his father tended not to intervene and in this sense was absent. I believe there were grievous consequences for James because there was no one present, externally and internally, to protect mother from him and to help him to be more separate from her. In this material I think we can see both the lure for him to snuggle up inside and his panic at being trapped there. The panic broke through when he heard the children calling "come inside," and there was no father there to stop him from getting inside. Kleinians believe the infant has an innate expectation of the penis and the primal scene, as well as of the breast. We think the baby recognizes the father when it meets him. When baby Sally was 9 days old, it seemed to me that she was already differentiating between mother and father, moving her head from the direction of one to the other, apparently by some recognition of their different pitch and tones of voice. From very early on we think the father provides an alternative object into whom the infant can project its loving and hating impulses. If the father is emotionally open and able to contain the baby's projections, then he may also lessen the impact of the baby's demands on the mother and so protect and support both her maternal reverie and her own psychic balance. When both parents, or the caregivers, are able to provide adequate containment for the baby by registering and giving meaning to its emotional experiences, then these can be introjected or re-introjected, and contribute to the formation and consolidation of good, dependable internal objects.

Depressive Position

A major developmental advance begins with the approach to the depressive position, starting around the third month. Aspects and functions of objects gradually come together in the baby's mind to form whole people: mother, father, siblings, and others. With this integration the baby starts to become aware that those who frustrate him are also those who love and satisfy him and at the same time comes a dawning awareness of a self who both loves and hates his objects. Ambivalence emerges. There is a shift in the focus of anxiety from survival of the self to concern for the object and anxiety about its loss or damage, and these fears give rise to depressive guilt, mourning, and reparation. Anxieties are intensified by weaning and by the baby's growing realization of its parents' separate relationship, which in turn stimulates envy, jealousy, and the pain of exclusion. Kleinians see the early Oedipus situation as coming to the fore at the same time as the depressive position is being established, and we think that 3 to 6 months is a time of enormous and complex development, with a rapidly increasing capacity to integrate experiences. Indeed, most parents seem to recognize a marked change in the baby during this time.

During this period there is an increase in introjective identification, partly due to the lessening of projective mechanisms and partly due to the baby's increasing awareness of his dependence on and separateness from his object, whom he can now perceive as a whole person. Anxiety about the loss of this object whom he loves and depends upon increases his need to possess that object and to keep it inside to preserve it from destructive impulses that, in phantasy, are projected out into "bad" objects. When good experiences with good objects are assimilated and benign qualities of these objects are identified with, the ego grows stronger and more differentiated. When the baby introjects its objects in the course of bad experiences, the ego feels an increase in anxiety and defends itself by splitting and projective identification. In the latter process a part of the ego is also projected into the object, thereby weakening the ego. Thus the ego oscillates between relative strength and weakness, between increasing integration and unintegration or disintegration.

Kleinians think it is crucial for future development that good experiences predominate over bad ones so that the infant builds up a benign and stable internal object that he loves and feels loves him, and upon which he feels he can depend. As the baby comes to recognize mother and father as whole people, he begins to realize that they are the source of both good and bad and that he, too, feels both good and bad toward them. This important change in perception allows the baby to begin to distinguish between phantasy and reality, between intrapsychic and external worlds. With the increase in perception he can test his phantasies with what actually happens and, given a reasonably benign cycle of experiences, his phantasies will increasingly draw closer to reality, both internal and external. For example, baby Sally's mother expected bossiness but was able to recognize a helpful midwife. The baby gradually comes to recognize his need for his objects and his helplessness without them. He also sees that his mother has relations with other people, especially with his father, and so he is exposed to feelings of jealousy as well as of envy.

Little by little the baby realizes that those he loves and depends upon can leave him. Anxiety about separation and loss increases his need to feel he can possess his good object, keep it inside, and protect it from his destructive impulses. His greatest fear, in the depressive position, is that he might destroy his entire world, internal and external. When he feels more integrated, he can remember his mother and feel sorrow when he feels she is lost to him. He may also feel guilt when he fears his attacks have damaged or destroyed her. We saw James's anxiety about the damage he felt he was doing in actually scraping the analyst's wall and, in phantasy, his mother's insides. When the baby feels he has lost his mother and he begins to experience guilt for his attacks on her, he resorts to massive defenses, such as projective identification, denial, or a manic form of reparation. He is exposed to despair, and the pain of this may drive him back to feeling surrounded by persecutors, the primitive persecutory superego. Then his internal world may disintegrate and fragment again.

In the paranoid-schizoid position, the primitive superego is shaped by the introjected idealized and persecutory objects and is experienced as ruthlessly demanding, punishing, and retaliatory. For example, as I suggested earlier, when the baby bites the nipple, the mother who flinches is experienced as biting back, so the baby freezes in defense. The ideal object is felt to demand perfection and in this exacting demand may also be felt as persecutory. In the early stages of the depressive position, the superego is still felt as severe and punishing. (I will discuss this further in regard to the adult material I shall present.) However, when relations with whole objects are more firmly established, the superego is gradually experienced as less monstrous and exacting, as not only the source of guilt but also as a loving source of support to the baby to withstand his destructive impulses. He gradually comes to wish to repair and restore what he feels he has destroyed, and these reparative wishes form the basis for later sublimation. It is in the depressive position that the baby begins to inhibit his impulses and anxieties and to displace them onto substitutes—the beginning of symbol formation (Segal 1964) and the foundation for the move forward to curiosity, intellectual development, mastery, and control. I want to emphasize, once again, that I am not suggesting Kleinians believe the baby *thinks* or *knows* what I have been describing. It is our hypothesis about how the baby's mind develops through experience of *its* experiences, as we discover in our analytic work with children and adults.

Clinical Material

I now wish to present material from an adult patient and to discuss this from the perspective of the movement between the two positions and the different levels of functioning. The patient is a single woman in her mid-fifties whose life has been blighted by the dominance of powerful unconscious phantasies, principally of a cruel, vindictive, and denigrating internal father and a weak, frightened internal mother who is experienced as unable to protect her and to whom the patient clings internally. Her actual parents have long been dead. The end of the analysis is now on the horizon. Following a period of improvement the patient had felt persecuted during sessions preceding the one I describe. She had felt that no one wanted her, that moves were afoot to sack her from her job, and she felt that I wanted to be rid of her as well.

She started the session in a querulous mood, describing that she had gotten nowhere, couldn't get on with the research paper she was writing, and felt so alone. She said her colleagues disliked her, she still hadn't had a reply from a senior colleague abroad, and she could not even reply to a letter from a male colleague (whom she likes). She said she was stuck, and she sounded hopeless. The situation and atmosphere conveyed a familiar mixture of hopelessness, provocation, and despair. I said that she felt that I did not want her back today when she felt stuck with feeling so lonely and unwanted, so she was anticipating a stuck session, expecting me to be both useless to her and to feel hopeless about her. She remained silent and motionless for a full two minutes. Then she wrapped the rug around her and curled up on her side in a fetal position in which she remained for the rest of the session.

After a while she spoke, in a complaining tone, of a meeting she and colleagues were to have with Mr. X., the head of their research institution. The meeting was to be about who was to succeed the retiring head of the department. As she talked of her anticipation of the meeting she sounded increasingly convinced that Mr. X.'s underlying purpose was to expose her in front of colleagues by indicating that she was not only not in the running but was to be made redundant. (The patient is, in fact, a senior member of the department.) Yet I was not sure that she entirely believed what she was saying, and I put this to her. Her response was to become more vehement about Mr. X.'s intentions. She concluded on a note of triumph that "he's a horrid little deathwatch beetle of a man, trying to get rid of me." Now she sounded as if she had convinced herself of what she was saying. I said I thought that she had felt I had not taken in the extent of her difficulty and the intensity of her feelings about her situation. While she did fear being eased out of her job, especially when she felt critical or had ideas of her own, she also believed that I did not want her either, especially when she felt taken over by deadly, vindictive, and triumphant feelings toward Mr. X. and me and, more important, toward herself. She continued to complain about her work colleagues, though less vigorously I thought, and she elaborated on details that became more complex and difficult to follow. I was now beginning to feel stuck and that I had lost contact with her. (I think this was the impact of her projective identification, which I discuss later.)

Eventually I said I thought we were now in a situation where she felt driven to pin me down with the complexity of her fears, both internal and external, so that I should feel I could not contemplate leaving her. If I were to consider such a thing, she believed I should feel very bad. She was silent for a while and then in a strangled voice said, "If I could only think that he doesn't want to get me out and if I could let colleagues have their say, maybe it could be a better meeting." I said she felt she now had these other thoughts: that if she could allow herself to think that I, too, might not be competitive with her, nor wish to be prematurely rid of her, then she feels it might be possible for us to consider her difficulties together, that is, to have a better meeting. After another pause, she said in a more reflective tone, "When I can't reach good relationships inside me, I can't write." I agreed, adding that she now felt there were better relationships inside and that she could let me know this.

Discussion

For much of the time in recent sessions the patient had felt herself to be in a persecutory world; she seemed to have started this session in a similar state. But from my long experience with her, the situation was a familiar one and it contained, I felt, an element of provocation. Yet the patient also commented on feeling stuck. I think that on her return she reestablished contact with me, externally as her analyst and internally with an object experienced as both demanding and resentful of her. The patient was relating in a characteristic way, not fully in the paranoid-schizoid position but stuck in what Steiner (1993) describes as a *psychic retreat*. Her retreat included perverse elements, "winding up" her objects, and hysterical features, clinging to her objects by exaggeration (Riesenberg-Malcolm 1996). Her initial response to the first interpretation was silence and stillness, rather like the baby who froze, then she curled up and covered herself with the rug, like the latency boy who went back "inside." I think she felt wrapped up at a more primitive level, and from there she became increasingly paranoid. It seemed that there was some exaggeration, as if she might want to draw me into persecuting her too, for instance by making a critical or exposing interpretation. When she said of Mr. X. that "he's a horrid little deathwatch beetle of a man, trying to get rid of me," I think this was a vivid description of what she wished to push me into being. It was her phantasy of a horrid, deadly presence watching her destroying herself as she acted out at work, and I was supposed to watch the deadly process and to feel helpless and hopeless. This was her provocation, and it would be her triumph.

In the countertransference later on, I had feelings of hopelessness and "stuckness," caught in the patient's web of complex detail. She had split off her experience of stuckness (which she had said she felt at the beginning of the session) and, in phantasy, projected this into me. The force and concreteness of this projective identification was such that it was some time before I could extricate myself internally suffi-

ciently to be able to think about what might be going on.
The patient characteristically has recourse to projective identification to rid herself of a part of her that feels helplessly tormented when she feels identified with a provocative internal
object. In the early part of the session I thought her provoca-
tiveness was an attempt to actualize (Sandler 1976) the phantasy by getting me to engage with her in this way. All would
agree on the importance for the analyst to recognize and
interpret, and not act out. One of the things that we find difficult to discern in our work, influenced as it is by our theory,
is the impact of projective identifications on the analyst.
These may be immediate, forceful, and concrete, or they may
be more subtle and cumulative (Joseph 1989).

In the transference the patient feared for the consequence
of what she was doing, that she would drive her objects to
reject her: in the external world, Mr. X., in the transference,
her analyst. That is, the analyst was, or would be, unable to
take in, in Bion's containing sense, her emotional predica-
ment, and so she feared that she would drive me into emotional withdrawal by her exaggeration of both the situation
and her emotional response. She may also have experienced
some persecutory guilt about using her analyst in this way.
The interpretation that she felt I had not taken in the extent
of her difficulty, or her feelings—especially vindictive, triumphant ones—seemed to help her to soften somewhat. The
next interpretation about her need to pin me down with the
complexity of her fears so I should not dare to think of leaving her, brought a significant response. She struggled to face
the possibility that she might think differently when she said,

"If I could only think that he doesn't want to get me out, and if I could let colleagues have their say, maybe it could be a better meeting." At that moment, when she softened and felt less persecuted, there was a shift in her perception. She became able to differentiate and to see external reality for itself. She made a move forward to further integration and a more depressive relation following the interpretation linking her new thoughts to me: that if she could think I might not be competitive with her, or wish to be prematurely rid of her, then we might have a better session. It was after this that she became insightful about the state of her internal world: "When I can't reach good relationships inside me, I can't write." She had become able to know her own psychic state.

Later in the Session

Suddenly there were sounds of motorbikes revving up outside, and the patient, obviously irritated, said that they were maddening. In brief, I linked the motorbikes to others coming to see me of whom she became aware, especially as it was nearing time for us to stop. I said that she felt provoked both by the intrusive sounds and by this thought that had come into her mind. In the last minutes of the session, after a silence during which I felt in contact with a more thoughtful patient, she said, "I'm suddenly seeing the Venice Opera House going up in flames" (she is fond of opera, and the burning down of the Opera House had been reported that morning). I replied that she sensed the end of the session, which she now experienced as me putting her down, and she felt her rage flaring up and threatening to consume us and the work we had done together today. After a long pause she said in a whimpering, plaintive voice that she didn't want to leave.

Further Discussion

We may contrast the adult patient's response with that of the latency boy when both experience sudden and unexpected external intrusion. James, in a regressed primitive state, enacting his phantasies of having damaged mother internally by his birth, reacted to the calls of the children outside with panic, disintegration, and confusion. He attempted to reestablish himself first by smelling and then using his brother's handkerchief, and then he went back "inside." That is, in phantasy he projected himself totally into his object, and he enacted this in the transference, in the session, by withdrawing into the couch. In this primitive infantile state, the senses respond: the smelling and holding on to brother's handkerchief. When the revving motorbikes burst into the session with the adult patient, she was feeling relatively integrated within herself and in good contact with the analyst. She also had more awareness, at that moment, of external reality, and so she was more able to tolerate the intrusion and the irritation provoked by it. When the patient felt in good contact with the analyst, she recognized her dependence on her. Then she struggled to face her resentment with the analyst who turns her out, as well as jealousy of the partner/patient with whom the analyst goes—in the transference and in external reality. Both her infantile and small self face jealousy of the internal parents' relationship with each other and with the siblings. I thought she recognized the siblings when she said if only she could let colleagues have their say. (In fact, the patient was the second of several children.) At this point in the session we see the patient approaching the depressive position.

On other occasions when she had felt more disturbed, precarious, or competitive, and outside noises intruded, she had gone into grinding resentment or paroxysms of rage, declaiming that I had failed to protect her or that I had even deliberately engineered the interference. When functioning at a paranoid-schizoid level, she would have felt the bikes and I were revving up against her, to get her out: both parents united against her infantile self. Alternatively, she might have projected onto the revving bikes her own hatred of the analyst who "kicked her out." Either way, she would have felt terribly persecuted.

It seemed to me that in the last minutes of the session the patient again felt in touch with her infantile self, so when she said—and note the present tense—she sees the opera house going up in flames, I think she was expressing her conscious fear of jealousy that brings out her rage toward her good object (the home of opera, of which she is fond) when the analyst ends the session. This includes her unconscious phantasy of what she deeply fears, that her rage fuels such destructiveness that she will destroy her internal world as well. She was more fully in the depressive position, and at this point she also faced the full impact of the early Oedipus complex. Her jealousy and envy drove her back to a clinging relationship with her object, and she ended the session as a plaintive little girl who doesn't want to leave. Did she feel driven to hang on in order to control her object from going to another, or to defend their relationship from her fury at being sent away, or both? I think that the pain of depressive anxiety and guilt was too much for her, so again she became the plaintive, resentful little girl who had had something taken away from her. The patient felt tortured by unconscious guilt at clinging to a long analysis, driven by her fear of giving it up and having to face the pain of its loss. For a woman in her mid-fifties endings also confront her with the realization of the last half of life and the pain of opportunities missed or destroyed earlier on.

Primitive experiences remain with us and feature in every analysis. Similarly, the constellations of the two positions, their anxieties, defenses, and characteristic object relations continue in mental life, unlike stages or phases of development that may be outgrown or worked through. However, Kleinians think the manner in which each position is established during the first six months is crucial for their subsequent reworking. If in the first months, when the paranoid-schizoid position predominates, the baby has a limited capacity to tolerate frustration (unlike baby Sally) and therefore resorts to excessive projective identification into a mother who, for whatever reason, lacks sufficient containment (or there is a mother who projects her anxieties into her infant, or both) then the ego does not strengthen, and the baby is more prone to states of fragmentation and confusion. In this situation the formation of a dependable good internal object is difficult and precarious. One outcome may then be that external relationships tend to be largely colored by hostile internal relations to objects.

Experiences of loss and relinquishment are a part of life, and whether they be great or small, they necessitate the reworking of depressive anxieties with respect to all previous losses. The pain of this psychic work may push the person back toward persecutory states or to resort to manic defenses. For example, the patient whose material I presented felt so terrified—both unconsciously and consciously—of facing the pain of being without analysis that she felt she could not bear to engage in the work of giving it up. To do so would also mean her having to face and work through the consequences of her destructive impulses, in phantasy and in reality, and so we had repeated and frequent negative therapeutic reactions. On the other hand, when the object is felt to be loving and good, and envy is not in ascendancy, reparative wishes emerge: the wish to restore in phantasy, and in actuality where possible, what is felt to have been spoiled or damaged. The success of each reworking of the depressive position depends upon the outcome of previous renegotiation, the resilience and flexibility of the ego, and the strength and support of loving internal objects. Love binds hate, and the process of reworking may lead to a deepening and enriching of the personality.

Implications for Practice

Kleinian understanding of infant development has important implications for technique. These primitive states of mind are relived in the transference, and it is there that they can be recognized and must be addressed in the moment when the patient is experiencing them and the analyst can understand them. Klein (1952b) wrote: "In the young infant's mind every external experience is interwoven with his phantasies and . . . every phantasy contains elements of actual experience, and it is only by analysing the transference situation to its depth that we are able to discover the past both in its realistic and phantastic aspects" (p. 54). As this dynamic primitive inner world becomes externalized through the analytic process, we try to trace and interpret its interactions in the transference. In this way they are emotionally relived and worked through. It is now widely held that interpretations should be about the interaction of patient and analyst at an intrapsychic level (O'Shaughnessy 1988). There are differences among British Kleinian analysts about whether and when we make links to the past. All, I think, would see the "reconstruction" of the past as occurring in the transference, where patients repeat their problems with their internal objects and we see the defenses they used in infancy and childhood. In this sense the present is more alive and real than the past, and we find it is often the patients who make the links to their past.

The language in which we talk to patients about their infantile aspects needs to be appropriate to their stage of development and level of understanding—both affective and intellectual. Thus, naturally, we speak differently to a 3 to 5 year old, a latency child, an adolescent, and an adult. Furthermore, we all tend to develop individual ways of talking to each patient, influenced by his or her and our own idiosyncratic manner of speech. Riesenberg-Malcolm (1986) suggests that interpretations should be expressed directly and concisely and that we try to describe in the present what is going on and why we think it is happening. Following Klein in the importance she attached to anxiety, we keep in mind the patient's dominant anxiety and whether we are interpreting affects, anxieties, or defenses. Another characteristic focus is the close attention we pay to the movement between the two positions and the response to interpretations. The clinical material that I have presented illustrates that focus.

References

Bion, W. R. (1957). Differentiation of the psychotic from the non-psychotic personalities. *International Journal of Psycho-Analysis* 40:308–315.

——— (1962). *Learning from Experience*. London: Heinemann.

Brenman Pick, I. (1992). Response to Daniel Stern's paper. *Bulletin of the British Psycho-Analytical Society* 28(6):29–30.

Cohn, J. F., and Tronick, E. Z. (1983). Three-month-old infant's reaction to simulated maternal depression. *Child Development* 54:185–193.

Field, T. M., Woodson, R., Greenberg, R., and Cohen, D. (1982). Discrimination and imitation of facial expressions by neonates. *Science* 218:179–181.

Fraiberg, S. (1987). Pathological defenses in infancy. In *Selected Writings of Selma Fraiberg*, pp. 183–203. Columbus, OH: Ohio State University Press.

Joseph, B. (1989). Projective identification: some clinical aspects. In *Psychic Equilibrium and Psychic Change: Selected Papers of Betty Joseph*, ed. M. Feldman and E. Bott Spillius, pp. 168–180. London: Routledge.

Klein, M. (1935). A contribution to the psychogenesis of manic-depressive states. In *The Writings of Melanie Klein*, vol. 1, pp. 262–289. London: Hogarth, 1975.

—— (1936). Weaning. In *The Writings of Melanie Klein*, vol. 1, pp. 290–305. London: Hogarth, 1975.

—— (1940). Mourning and its relation to manic-depressive states. *International Journal of Psycho-Analysis* 21:125–153.

—— (1946). Notes on some schizoid mechanisms. In *The Writings of Melanie Klein*, vol. 3, pp. 1–24. London: Hogarth, 1975.

—— (1952a). Some theoretical conclusions regarding the emotional life of the infant. In *The Writings of Melanie Klein*, vol. 3, pp. 61–93. London: Hogarth, 1975.

—— (1952b). The origins of transference. In *The Writings of Melanie Klein*, vol. 3, pp. 48–56. London: Hogarth, 1975.

MacFarlane, A. (1975). Parent–infant interaction. Ciba Foundation Symposium, 33, pp. 103–117. Amsterdam: Elsevier.

Murray, L., and Trevarthen, C. (1985). Emotional interactions between two-month-olds and their mothers. In *Social Perception in Infants*, ed. T. M. Field and N. A. Fox, pp. 177–197. Norwood, NJ: Ablex.

O'Shaughnessy, E. (1988). Words and working through. *International Journal of Psycho-Analysis* 13:73–89.

Piontelli, A. (1992). *From Fetus to Child.* London/New York: Tavistock/Routledge.

Riesenberg-Malcolm, R. (1986). Interpretation: the past in the present. *International Review of Psycho-Analysis* 13:73–89.

—— (1996). How can we know the dancer from the dance? *International Journal of Psycho-Analysis* 77:679–688.

Rosenfeld, H. R. (1950). Notes on the psychopathology of confusional states in chronic schizophrenia. *International Journal of Psycho-Analysis* 31:52–62.

Sandler, J. (1976). Countertransference and role-responsiveness. *International Review of Psycho-Analysis* 3:43–47.

Segal, H. (1964). The depressive position. In *Introduction to the Work of Melanie Klein*, pp. 67–81. London: Hogarth.

Steiner, J. (1993). Perverse relationships in pathological organisations. In *Psychic Retreats*, pp. 103–115. London: Routledge.

Stern, D. N. (1985). *The Interpersonal World of the Infant.* New York: Basic Books.

—— (1991). Consideration of unconscious phantasy from a developmental perspective. *Bulletin of the British Psycho-Analytical Society* 29(2):2–11.

—— (1993). One way to build a clinically relevant baby. *Bulletin of the British Psycho-Analytical Society* 27(13):2–10.

6

Early Development and Disorders of Internalization

Arnold Wilson and
Julia L. Prillaman

Introduction

In this chapter, we aim to elevate into conceptual significance the clinical importance of internalization so that it is squarely brought forward as an aspect of what psychoanalysts treat.

Internalization deserves more careful clinical scrutiny in the current world of contemporary structural theory. Internalization encompasses cognitive understanding as well as underlying psychic structure formation. A deeper understanding of this important component of a therapeutic analysis can help shed light on key processes underlying the progression of a treatment. This chapter will include the concept of internalization and the clinical manifestations of what we will term *internalization disorders*. In examining internalization, we will link data from the treatment setting with extra-analytic data, emphasizing data derived from empirical research and using it to inform theory construction. It is through internalization that an analysand is empowered to make use of the clinical setting. Thus the processes underlying the assimilation of meaningful experiences and insights by an analysand sets the backdrop for a successful analysis.

Although we utilize constructs from outside traditional analytic theorizing, our approach in discussing internalization disorders falls within the broad scope of contemporary Freudian or "structural" analysis, because our primary emphasis is on intrapsychic structural considerations. Some of our proposals can be thought of as within the realm of what Freud (1915) described as the descriptive rather than

dynamic unconscious. More specifically, we address internalization disorders as structured processes ingrained within an
underlying layer of mentation motivating habitual actions, or,
as Frank (1969) aptly put it, the "unrememberable yet unforgettable." Disorders of internalization are subject to repetition, and we will explore how this is so.

The Perspectives of "Differentiated Selves" and "Nondifferentiated Selves"

The limits imposed upon a person's perspective largely
determine what that person sees. In this day and age of comparative psychoanalysis, no one perspective yet articulated
allows us to see everything. It is important to keep in mind
that different angles can yield different observational data,
each of which can make a distinct claim of "objectivity." In
science, it is enormously difficult to pin down how theory
and data determine a concept of "objectivity" without careful
consideration of a larger domain of evidence than at first
glance is perceived. We propose two root metaphors (Pepper
1942) that are currently in conflict in most contemporary
theory construction. Both form and reform data, and when
extended, imply technical considerations for the psychoanalyst. Both can lay claim to an overarching status, although in
fact neither achieves this distinction. One is the *attached dyad*
and the other is the *differentiated dyad*.

We propose a specific view of internalization disorders
from the perspective of the poorly differentiated dyadic
matrix—and we propose an analogy between the caregiver–

child and the analyst–patient. This perspective yields remarkably different data than the perspective of a matrix of two separate, psychologically sentient beings. From the first perspective, minds are shared and interdependent; from the second, minds are separate. The distinction between these two perspectives is important because internalization disorders swim more into focus when viewed from the perspective of analyst and analysand as attached beings. The human interactive context has a remarkable effect upon how, why, and when internalization takes place, much of which is outside the scope of verbal exchanges. The view from a differentiated angle lends itself to an emphasis upon interpretation of ideational conflict, in all its diversity of expressions. It makes much involvement often appear to be an error, at times silly. The view from the attached angle lends itself to an emphasis upon the analyst's involvement in the back-and-forth flow of events, in order to utilize what is currently happening as a source of data. It makes interpretation often appear to be an error, at times silly. Reconciling tensions between these two seemingly contradictory tasks can be thought of as defining a beneficial analytic process, as argued in a recent paper (Wilson and Weinstein 1996). A tilt toward one side or the other of these seemingly disparate technical approaches may well lead to more-or-less large sectors of an analysand's organization being overlooked, depending, of course, on the particular organization of the analysand, and, of course, assuming that the resulting analysis is viable. We tend not to learn from our mistakes because the recipients are not around to haunt us. It is our impression that most analyses can usefully benefit from a recognition that both are at play. It is from the perspective of observations on naturally occurring dyads, that is, caregivers and children, not springing from the analytic setting, that the angle on attachment is elevated into prominence and we are prepared for the inferential leap into first recognizing disorders of internalization and then treating them in the analytic setting. This privileging of data from outside the clinical situation confers a different kind of objectivity on the data than upon clinically derived data; for acceptance, one passes controlled tests of a priori probability, the other undergoes less rigorously controlled tests of efficacy in an ecology better suited to their validity. We submit that they are both complementary and provide for hypotheses that can each be tested in the other's arena.

Viewing internalization disorders from an attached perspective thus affords an opportunity to recognize distinct aspects of the psychoanalytic process (i.e., the reciprocal nature of the analyst and analysand's contact) in a different way than when viewing them from the differentiated perspective. Conceiving of the cyclic flow of analytic interaction from an attached context helps clarify why some of the recent attention paid to the here-and-now transference is theoretically sound.

Both perspectives, however, are necessary for psychoanalysis to fully conceptualize the range of phenomena within its field. Although each provides the illusion of carving out a sound and comprehensive view of psychoanalytic technique, these two perspectives do not have to conflict; they can comfortably coexist. It is a task of contemporary theory construction to promote this coexistence. Internalization disorders, therefore, can be defined as comprised of processes preventing an analysand from learning from the analysis—within the unique social ecology afforded by the analytic setting. They presuppose that two people are invariably separate in some ways yet irrevocably attached in others, but that the interactive flow of events yields crucial information concerning how, when, and why internalization takes place.

In conceptualizing the dyadic basis of certain experience, some analysts take quite literally a comparison of parent–child with that of the analyst–patient, and end up stipulating exact parallels between caregivers and children with analysts and patients. We suggest value in the comparison only when not applied in too concrete and isomorphic a fashion, as when suggesting that the analyst raises the patient much as a caregiver raises a child (see Mayes and Spence 1994). This reifies the comparison and in doing so sacrifices analytic explanatory power. Concretization of the analogy can lead to aberrations of the tasks of analysis, such as the transference cure, or where at the end of an analysis the patient's identification remains with the analyst rather than the analyst's analyzing function. Yet another aberration is how the overriding value of linking words to previously unorganized experience may be lost.

Early Developmental Processes Impacting upon One's Ability to Internalize

As stated earlier, interactive processes play a large role in the subtle ways internalization takes place. Internalization disorders are complex and multidetermined, and many factors may contribute to their development. However, we will focus on three sets of phenomena as potentially influencing the development of internalization difficulties: self-regulation, the developmental principle of epigenesis, and language acquisition. Each of these processes is thought to significantly contribute to the analysand's ultimate ability to internalize the analytic work, although they certainly do not give the whole early picture. A multifaceted approach is necessary in order to fully grasp how internalization does or does not take place. Our emphasis on these particular phenomena attempts not to reduce internalization disorders to three specific causes but, rather, to avoid confusion and the inevitable cursory approach that would attend an exhaustive list of contributing factors. We begin our examination with the concept of self-regulation.

Self-regulation

Wilson and colleagues (1990) define failures of self-regulation as an individual's inability to independently control and contain key internal processes. As a result, the individual uses an external other's influence to help control and contain himself. The other becomes essential for the smooth continuity of a wide medley of psychological and biological processes. When one person plays a functional role in the determination of another's ability to regulate key psychological processes, it constitutes a failure of self-regulation. This use of an external other can take many subtle forms. Hofer

(1984) vividly documents such subtle forms in his discussion of the term *hidden regulators* in the lives of children. To add a psychoanalytic spin to Hofer's findings, we assert that behavioral regulatory patterns exist in which relational objects serve as regulators. Hofer (1984) asks: "Could the elements of the inner life that we experience with people that are close to us come to serve as biological regulators, much the way the actual sensorimotor interactions with the mother act for the infant animal in our experiments? And could this link internal object relations to biological systems?" (p. 15).

In answering this question, Hofer demonstrates mutuality of regulation. The symbiotic attachment incorporates a myriad of hidden processes that the mother regulates in the child (e.g., activation patterns) and the child regulates in the mother (e.g., lactation, arousal). When the attachment system is working fluently, these processes remain quiet and hidden under the umbrella of symbiosis. When attachment is disrupted because of untoward separation, these processes can flare into visibility as discrete and distinct processes in conflict. For example, the sleep/wake cycle can become disjointed from other behavioral and physiological patterns within the infant or within the mother, at which time the infant can experience a failure of self-regulation with potentially pathogenic consequences. The human core formed by an enduring symbiosis is one marked by development of the ability to self-regulate and manage stimulation levels.

Self-regulation's importance is highlighted by the fact that, when intermittent or chronic failures of self-regulatory processes occur, external regulation sources are sought to perform the same psychobiological task. Recruiting others for regulatory purposes is normative and not pathological in children; its status as normative or pathogenic depends upon the timing of its emergence. The baby who symbiotically fuses with its mother for regulating emotional arousal is behaving normatively; the adult who symbiotically fuses with another adult for the same purpose is behaving in ways attributable to schizophrenia. Thus, within the jurisdiction of early internalized object relations, the strength of one's constitution is influenced by patterns of symbiotic recruitment of objects. If recruitment leads to successful regulation, the child is then able to internalize the parent's interventions, ultimately leading to the development of independent regulatory functioning. If the parental figure is remiss in his/her ability to aid in the development of the child's regulatory functioning, the child internalizes either a consistent unhealthy regulatory process or fails to internalize any particular regulatory pattern, leading to regulatory chaos. A breakdown in self-regulatory development, therefore, negatively affects internalization capacities, thereby necessitating greater analytic emphasis on promoting these capacities. Anna Freud (1970) demonstrated how difficulties in the pain/pleasure management during the first year of life can preempt developmental achievements. She then focused analytic work with children of this age upon regulation failures.

Self-regulatory failures manifest themselves over an extended course of time. Radical changes in one's average expectable environment (such as serious medical illness or sudden traumatic loss) can result in unmanageable anxiety or

panic states or sleep disorders in people with no apparent previous psychopathology. Otherwise healthy patients who suf-
fer acute trauma are quite different from those who exhibit
ongoing characterological disturbances that are a chronic
manifestation of self-regulatory failure. All in all, over a life
span, failures of self-regulation can lead to a multitude of psychological and psychobiological difficulties, both acute and
chronic (Emde 1988), and encompassing both symptom and
character. Thus, failures of self-regulation have been theoretically tied to a variety of psychopathologies, including narcissistic personality (Wilson 1989), severe characterological
opiate addiction (Khantzian 1978), and borderline and
schizophrenic conditions (Grotstein 1983, 1989).

Self-regulation is part of a series of constructs that empirically lend evidence to support the claim that the original pat-
terning of the mind cannot be adequately conceptualized as
bounded or contained within individuals. It is later that the
mind becomes "privatized" (Wilson and Weinstein 1992b)
and both versions of the mind coexist within the structural
organization of the adult psyche. To borrow Wertsch's
(1990) phrase, the early mind "extends beyond the skin"; it
is a dyadically forged entity that eventually becomes localized
within the individual after extensive psychological development has taken place. The progression from other- to
self-regulation parallels the development of a separate mind.
A good deal of research indicates the need to conceptualize
this progression as inherently bi-directional. Just as the other
regulates the self, the self likewise regulates the other.
Self-regulation implies a nonlinear, multiply determined,

dynamic, and gradual acquisition of selfhood rather than one that is reducible to single factors.

Epigenesis

A second contributing factor to the development of an internalization disorder is a breakdown in the normative epigenetic process. In contrast to a stage theory, epigenesis is organized explicitly around both continuities and discontinuities in development. An explanation of epigenesis will help connect it to internalization disorders. Epigenesis is a meta-theory of development specifying the way in which transformations occur between levels of developmental organization. Theorists have used epigenesis to study models of development in the psychological sciences to conceptualize a superordinate principle that explains the movement from psychobiological to more explicitly psychological functioning (Kitchener 1978, Wilson and Passik 1993). To briefly summarize, the biological view of epigenesis has five distinct tenets. (1) A causal sequence of interactions exists between organism and environment. (2) Earlier and more undifferentiated structures in the organism interact with the environment and cause more differentiated structures to unfold in the organism. (3) These structures unfold in a series of levels, stages, or modes. (4) these structures come to possess increased levels of complexity, differentiation, and organization. (5) Particular emergent qualities characterize each new level. Thus each particular level or mode is characterized by a discernible degree of complexity and differentiation of structure. Each mode possesses certain emergent qualities, denotes a particular interplay between organism and environment, and causally relates to all past and future modes through the nature of successive transactions that reorganize structure and experience. This leads to a primary focus on the transformational properties and principles of structure and a secondary focus on content, or the phenomena that is undergoing transformation. One advantage of this approach is that we can understand changes in mental life that have both qualitatively continuous and discontinuous qualities.

For the purposes of psychoanalysis, one must extrapolate from biology to find concrete application of this metatheory of development. We adduce several additional derivative postulates that characterize the importance of epigenesis for psychoanalytic theory in particular. First, the formation of psychic structure is the result of successive reorganizing transactions between the child and the caregiving environment, each of which can be termed a *mode*. Next, the form or structuralization of each mode of organization depends upon the outcome of each previous mode and the subsequent effects of experience. Each mode integrates previous modes and results in new and more differentiated and articulated levels of organization and regulation, and each mode is defined by its own emergent properties. Furthermore, there are two important corollaries to these postulates: (1) once a given mode has been integrated by a higher mode, the more archaic form nonetheless remains as a potential endpoint of regression, at which time it can become the overriding organizer of experience; (2) lower modes are contained within and can continue to exert influence over higher-order, advanced psychological processes despite their having yielded to developmental transformation.

These transformational principles add a remarkable complexity to how analysts might model the processes of internalization, taking us far beyond psychosexual and cognitive metaphors and most "stage" models. Epigenesis provides a potential blueprint for the ontogenesis of mind in relation to both endogenous and exogenous forces. Principles for understanding the dynamic flux over time emerge as crucial and complement the traditional understanding of the contents of the mind. Dimensions of psychic structure can be seen as fluid and multileveled, subject to more or less constant regressive and progressive influences. Regression and progression are constant aspects of clinical reality, as a person can access structures at virtually any level and at any moment. Through regression and progression, any developmental form of the mind becomes accessible and must be accounted for by the analyst. Any pattern of habitual action becomes accessible under the tax of regressive influence.

The nature of the mind is now of a different sort than analysts customarily depict—something far more mutable and momentarily reactive. This stance on psychic structure departs from the definition by Rapaport and Gill (1959) and Kernberg (1976) that structure is stable and slowly changing. This stance supports Loewald's (1960) view, however, that technically one of the analyst's most important tasks is to cartwheel along with the analysand through volatile cycles of regression and progression, always tuning into the prevailing state, always offering the analysand the sense that regression can be met and that higher states are available and can be realized. In this way, a new and less deceiving reality beckons. Through psychoanalysis one spirals forward and upward as one works through conflicts and internalizes insights. This movement forward is thwarted if the ability to internalize is compromised. The analysand, if limited in his ability to internalize, is unable to reach higher developmental levels. Thus it is imperative that an analyst consistently and carefully examine an analysand's capacity to internalize throughout the analysis so that decisions can be made about *transference and counter-transference positioning* (a term to be addressed shortly).

Language Acquisition

A third factor in the development of internalization disorders is language acquisition. The pragmatics of language play a crucial role in the psychological capacities that promote internalization. Language allows persons to objectify and symbolize their relationship with the world, thereby aiding in the regulation and internalization of ongoing experience (Wilson and Weinstein 1992a,b). Further, the emphasis upon language can have a subtle impact upon one's scientific world view. For example, Cavell (1993) has aptly noted that an emphasis upon language can render obsolete the distinction between intrapsychic and interpersonal. Cavell (1989a,b) distinguishes between Cartesian and interpersonal philosophical models of science and mind. She uses the term *interpersonal* to refer to a fundamental world view that encompasses language and affect as vehicles through which the primary communicativeness of human nature is expressed, not as reference to any interpersonal school of thought in psychoanalysis (e.g., those of Sullivan or Horney). She emphasizes a trend in recent philosophical models to refer to how the phenomenon of meaning is dependent upon an interaction between minds that is built into nature—there is no natural communicational chasm between people that must be overcome in the interpersonal models, as is the case in what she terms the Cartesian models. The interpersonal concept of mind has organic logical links to language as a medium through which minds are indissolubly and nonsolipsistically linked. Philosophers associated with this recent trend include the "late" Ludwig Wittgenstein and Donald Davidson.

It is no accident that Greenacre (1968) hypothesized that those phenomena native to the early stage of language acquisition block key skills necessary for an adult analysand's successful, ongoing analytic process and that therefore cause certain analysands who initially appear analyzable to ultimately fail on the couch. Greenacre states:

> For the extent to which the interpretation can be made assimilable to the analysand depends not only on the analyst's sensitivity to the content of the patient's transference productions and his adequate knowledge of technique and principles but further on the construction—the stuff of which the patient's early attachments and identifications have been made in the period up to and including the acquisition of speech and the time immediately afterward. [p. 769]

Following Greenacre's lead, we will briefly identify work on language acquisition that pertains to self-regulation and the interactive, mutually regulating, social and cultural domain from which language emerges. Then, we will offer some speculations on the concept of a linguistic representation as one way analytic theorists might map the mind in the future. Bruner (1983), like Vygotsky (1988a), speaks of a child's language as "the means for interpreting and regulating the culture. The interpreting and negotiating start the moment the infant enters the human scene. It is at this stage of interpretation and negotiation that language acquisition is acted out" (p. 24). Vygotsky viewed the progression from other-regulation to self-regulation as occurring not only within the mother–infant dyad. In his view self-regulation becomes a general learning principle that occurs throughout life in interactions with important others, embedded within the jurisdiction of culture at large. His analysis of the movement from other- to self-regulation is similar to Schafer's (1968) more familiar psychoanalytic definition of internalization as being those processes by which the subject transforms real or imagined regulatory interactions with his environment, and real or imagined characteristics of his environment, into inner regulations and characteristics (p. 9).

Vygotsky introduced the phrase *zone of proximal development* to describe the distance between the actual developmental level (what skills one has internalized) and the potential developmental level (what competence one can realistically reach with another's aid). Early social objects act virtually as prosthetic devices in the creation of the child's mind. Vygotsky was one of the first psychologists interested in the ontogenesis of mind to arrive at the valuable formulation of the ways in which early dyadic interactions become internalized and made into psychic structures. Through the auspices of language, two individuals co-construct internal states that come to constitute the child's mind. In Vygotsky's view, the mind is actively created in the interchange within the zone of proximal development, as crafted by the progressive use of language by both child and caregiver. Their intersection of intention, as mirrored in language, is the inevitable context or unit of study, not any aspect of the child in isolation. This point is important, because to Vygotsky the "word meaning" is the primary unit in studying human nature. In sizing up this unit of analysis, Wertsch (1985) argues that the tool-mediated and goal-directed action rather than the word meaning should be the basic unit of analysis in Vygotsky's system. This is because it is such an action that best reflects the interfunctionality that stamps Vygotsky's system, whereas word meaning is a unit of semiotic mediation rather than mental functioning itself and therefore is too limited. We do not follow Wertsch in this revision because we think that to do so leads to the loss of something central to Vygotsky's system, namely the genetic emphasis on the intralinguistic context of word meanings and their flux over time and situation. Nevertheless, Wertsch's argument that such actions are more tenable than word meanings as central units of analysis because they partake in more mental functions and so lend themselves to actual concrete psychological research is certainly logical. For a detailed discussion of this issue, see Zinchenko (1985).

Vygotsky's (1988a) views on the developmental intertwining of language and thought are also important. He traces the development of thinking and phonetic production from their most rudimentary forms, until they reach the point at which, through their intertwining, they become necessarily and permanently transformed. It is interesting that these two move in opposite directions along the course of their development. At first, a very small phonetic unit encapsulates a large constellation of thought. As language acquisition proceeds, phonetic production expands, while the thought expressed with each word diminishes. It then takes more words to express the thought, but only then can that thought be fully differentiated, because it has been articulated.

It is important to understand that Vygotsky's theory of language acquisition is in no sense an associationist theory. Words take on significance in an interpersonal context and because of their functional value. For Vygotsky (1988a), this functional capacity of the interpersonal use of language is what can potentially become internalized and serve the same function in the intrapersonal sphere, through the development of what he terms *inner speech*. Thus the plane of interaction of language and thought is social intercourse, and the glue that unites them is word meaning. Therefore, meaning is created by the unification of speech and thought. This is one example of the way in which language and thought coexist in a reciprocal relationship. This also demonstrates that the changes that language and thought go through are not quantitative, but qualitative. It becomes apparent that this approach to psychological development has the potential to describe the development of many other aspects of affective/cognitive functioning, which are both reflected in and partially determined by language and the interpersonal history of its acquisition. These phenomena are well documented by psychoanalysts, who have not previously had a sufficient theoretical framework within which to fit them.

Central to Vygotsky's (1988a) model of linguistic development is the concept of "interiorization," which in his analysis consists of the interiorization of speech. He traces this process through the functions of speech, which he sees as traveling down two distinct paths from a common starting point. To Vygotsky, all speech originates as social speech. Its purpose is for interpersonal communication. One of the first functions of the infant–caregiver dialogue is that of regulation of the child's behavior by the caregiver. The next major stage of development involves the adoption of this strategy by the infant, who begins to use language to influence the behavior of others. At this point there is a functional split in language. One path it follows is the increasing development of social speech. The other path takes speech inwards as "speech for self." The child takes the speech that has been learned as capable of controlling others and uses it for control, as in the example of the child who says "no" in order to inhibit action. Vygotsky was able to trace the forms that this speech took by following this functional trail. He analyzed the egocentric speech, first described by Piaget as failed social speech. Contrary to this depiction, Vygotsky recognized that as part of normal development, the child's speech becomes more egocentric until it disappears altogether. He studied the characteristics of this speech, and noticed that it was characterized by greater predication (truncation). It is as if speech comes to serve self-regulation more effectively so that it needs to be less fully articulated, until it eventually becomes silent, that is, inner speech. Vygotsky also demonstrated that this speech for self could be brought out again under conditions of stress, or when obstacles are placed in the path of action.

This view of language acquisition and its implications for the mind was originally not influenced by psychoanalytic thought. We do know that Vygotsky read and was alternately disparaging and intrigued by the works of Freud, but did not make Freud a central figure in his publications (see Vygotsky 1988b,c). Psychoanalysis itself in Russia had early adherents, but the theory was soon jettisoned because it lacked the social foundation necessary to mesh with the developing Soviet ideology. Luria (1979), in his autobiography *The Making of Mind*, depicts how he founded the first psychoanalytic association in Russia before he met Vygotsky. This psychoanalytic association did not last very long, for a variety of reasons. Luria and the linguist and literary critic Michael Bakhtin both had early flirtations (of which Vygotsky tended to be critical) with Freudian psychoanalysis; Luria corresponded with Freud and had an editorial involvement with the early *International Journal of Psycho-Analysis*. The rationale for the Soviets' rejection of Freud and the American embrace of the solitary child now amusingly appear to have run full circle and reunited, and psychoanalysts in the United States have arrived at an exciting new interactive theory of childhood with a strong flavor of dialectics, while the Soviets have themselves recently rediscovered Freud, according to the *New York Times* (July 18, 1988), as a "visionary . . . [who can] offer Soviet psychiatry tools it badly needs."

It was suggested that psychoanalysis might look toward mapping the mind with units of language (i.e., of linguistic representation). This task is well beyond the scope of this chapter. However, the concept that linguistic representability might meet some of the requirements for a broader conceptualization of representation is not new to psychoanalysis. At a panel on the relationship between language and the development of the ego, the following question was asked: "To what extent is the organization of the ego reflected in the language system, and to what extent does the internalized language system constitute a supraordinate regulating system of the ego?" (Edelheit 1968, p. 114). The suggestion was "the process by which the normal child acquires the complex patterns of speech suggests a linguistic model for the development of psychic structure" (Edelheit 1968, p. 115).

Yet, in this panel, the first provocative question is not satisfactorily answered, and the second bold suggestion dangles in the absence of a theoretical approach that would permit further explication. Unlike many other fields of inquiry concerned with understanding the workings of the mind (see Shapiro 1979), psychoanalysis does not make a provision for a linguistic form of inner representations. Following Freud's

views on word and thing presentations, psychoanalysis histor-
ically limits its assignment of the function of language to promoting the reorganization of unconscious impulses into
communicative form providing access to consciousness.

The idea of representation implies some inner facsimile,
duplication, some isomorphism with the outer world, albeit
one subject to influence by prevailing fantasies and affect
states. Inherent to representability is an assumption to which
we want to draw attention. It is often assumed that psychoanalysis requires a representational world composed of units
that can alone or in combination stand for the external world.
A representational world must then be composed of those
units that allow the mind to offer a "mirror of nature," to use
Rorty's (1979) apt phrase. To the extent that psychoanalysis
persists in viewing the child as having the sole supraordinate
task of knowing the real, and then representing a copy of the
real in order to continue one's involvement with it, then the
representational world is by necessity conceived of as built of
the bricks of such units of knowing. This usually takes one
form or another of the image representation (Friedman 1978).
We do not agree. One might, for example, opt for the Witt-
gensteinian notion of language as a tool that engages the
world, rather than serving to set the stage for a mirror of the
real. When the mind is viewed this way, the rationale for a
linguistic representation then appears as a possibility. One
can hence view another supraordinate task of the child as grap-
pling with the world as well as copying it. For example, based
on data from the clinical situation, Arlow (1969) proposes
that unconscious fantasies must contain an imagistic or picto-

rial component as well as a linguistic component. Thus, the linguistic representation becomes what Goodman (1984) terms a natural referential unit of the representational world. To Goodman, *reference* is a "primitive term covering all sorts of symbolization, all cases of 'standing for'" (p. 55). It is this shift of fundamental assumptions upon which we rest our proposal of the possibility of a linguistic representation. Such representations can be gauged not only by their stability but also by how well they achieve communicability. These representations are not determined solely by input but also by their bidirectional engagement with other inner and outer objects.

Notice that one assumption and its implications are not more "true" than the other—both follow from different assumptive approaches to psychoanalytic theory. It is the former view, of atomistic units that can comprise mind which additively summate to meaningfulness, that Wittgenstein (1953) mocks in *Philosophical Investigations*. Some post-logical positivist philosophers of science have termed similar views *naive realism* and described them at length as untenable in the modern world (Bernstein 1984, Hacking 1983, Hesse 1980). Geertz's intent (1973), in his discussion of "thick" description, attempts to overcome cultural "formism" and to point cultural anthropology toward a similar, more interpretive stance.

We then are led to ask: Is the theory of language acquisition and of linguistic representation in a separate domain from psychoanalysis because psychoanalysis does not acknowledge inner speech as representational in its own right? Much stands to be gained in psychoanalysis by allowing for linguistic representability. We propose that the dialogic nature of the theory of early development presupposes a concept of inner object representation, thus requiring the postulate of an introject. Vygotsky's "speech for self" implies the presence of an internal object. The dialogic nature of speech assumes two intrapsychic representations, in conversation with each other in some way. There are striking similarities between our construal and Loewenstein's (1956) description of what takes place during self-analysis. Loewenstein describes how inner speech plays a central role in the therapeutic action of psychoanalysis outside of the clinical situation. He notes that "its effectiveness . . . [is] in the form of a continuation of a previous analysis with an actual analyst . . . it is then usually a solitary continuation of dialogue with the latter or with an imaginary analyst . . . it might be viewed as an imaginary dialogue in which the subject is able to play both parts, that of a patient and that of an analyst, and thus to some extent involving inner speech" (p. 467).

The importance of internal dialogue is further highlighted by pointing to evidence suggesting that it is through the inner dialogue of the self with the internal object, spoken in the phonetically silent tongue of inner speech, that a child is able to gain access to other crucial component processes of self-regulation (Wilson et al. 1990).

What of inner speech as one among other forms of representability? What the metapsychological status of language is, how it lingers on within the mind as psychic structure, and how it fits with other macrostructures and microstructures, are crucial questions. We do not think developmental psycholinguistics lends itself easily to being transported into or grafted onto psychoanalysis. Nor can we simply hypothesize an unconscious Chomsky-like language generator that is postulated at the same metapsychological level of abstraction (Waelder 1962) as other psychic structures. This metapsychological coordination is a daunting project.

For now, we merely note that there are some precedents for psychoanalysis to consider inner speech as a form of representation, from inside and outside psychoanalysis. For example, Isakower (1939) suggests the auditory sphere has a preeminent place in the formation of the superego, and then the superego becomes experienced as a form of language above any other form of presentation. Horowitz (1972) picks up and explores the implications for psychoanalytic theory of Bruner's (1964) hypothesis of representations of enactment, image, and language, each as distinct levels of representation. Translation from one level of representation to another becomes one of the most important overarching developmental problems, literally defining the essence of defense to Horowitz. From the field of cognitive science, Johnson-Laird (1983) also develops a triarchic theory of levels of representability. The status of a linguistic form of representation is currently quite controversial in cognitive science (see for example Marslen-Wilson 1989). In his contribution to a fury of contention over this topic, Johnson-Laird notes that there is experimental evidence that indeed a useful distinction can be drawn among representations as mental models which are structural analogues of the world, images which are the perceptual correlates of mental models, and propositional representations which are symbols that correspond to natural languages.

The concept of intersubjectivity has recently arisen in American psychoanalysis. Rommetveit (1974) defines intersubjectivity linguistically. It is, in our view, most useful for psychoanalysis to define intersubjectivity according to linguistic communications theory (e.g., Rommetveit 1974), as when two or more interlocutors share some aspect of their situation definitions in the negotiation of sense. With this definition in hand, intersubjectivity can become an important concept not heretofore present in the psychoanalytic literature. It assumes the possibility of a sharing of minds without wrenching the two minds into Descartian assumptions of subject and object. We do not endorse the definition given by Stolorow and colleagues (1987), who define intersubjectivity as a field within which two subjectivities meet. This definition is so general as to stipulate virtually nothing and therefore is not useful except as a polemic. We also see problems with the well-known definition from developmental psychology of Trevarthen (1979), who defines intersubjectivity as a particular developmental stage. This radically limits the term for analysts and makes it virtually inapplicable to adulthood, unless intersubjectivity in adulthood is understood as requiring a regression to infancy.

Making clinical sense of the ideas and research on language acquisition and use requires a shift in context. The dynamic context within which language is acquired must be understood within the nexus of primary object relationships—their anxieties and conflicts. This is why Stern (1985) points out that wishing, defensive activity, as well as the deep-est symbiotic attachment between infant and caregiver are conceptually plausible only with the acquisition of language capacities. The acquisition of language occurs in the context of the conflicts of early childhood (e.g., fear of loss of love and the object, moral anxiety, or fragmentation of the self), and must reflect the ways in which such conflicts are defended against. Thus, rather than being an example of conflict-free functioning or a sheer cognitive skill, language acquisition reflects the same psychodynamics of early conflicted object relations as do such frequently studied contemporaneous phenomena as separation or gender identity consolidation.

An extensive body of psychological knowledge can be found about how internalization takes place and implications can be drawn from self-regulation, epigenesis, and language acquisition. Each of the three domains informs elements of early communication that can lead to a failure in establishing the object as a separate entity, and that therefore becomes drawn in so that a disorder of internalization becomes present. This, then, can usefully be distinguished from a disorder of conflict, in which explicitly intrapsychic considerations pre-dominate. Turning to the analytic situation, one can say that the failure to take in from the analyst coexists with intrapsychic conflict within the scope of clinical psychoanalysis. Disorders of internalization take many subtle forms, often at the inter-stices of verbal content rather than semantically organized by the brain, and can be treated using the analytic method. Inter-nalization disorders are most visible, made manifest, in the so-called "here-and-now" interaction. Historical consider-

ations are of some intellectual interest but are largely superflu-
ous given the immediate urgency of ongoing processes. They
can be found not so often in the content of an analysand's verbal productions, but rather in the actions an analysand
embarks upon within the analytic setting, where transference
and countertransference positioning are important.

Transference, Countertransference, and the Analyst's Optimal "Positioning"

To begin with, in an optimal analysis, the analyst and
analysand co-construct the psychic states necessary to move
the analysis forward, while setting about analyzing the needs
and wishes emanating as much as possible from the analysand
that comprise the transference. The traditional definition of
transference is the externalization of unconscious fantasy
onto the inner representation of an ambiguously perceived
object—not comprising any and all interaction between the
analyst and analysand. To co-construct certain mental states
then allows the transference to the differentiated object not
to be co-constructed. In other words, it is the zone of proximal development that undergoes co-construction, but not
the transference. This leads to the conclusion that an analyst
strives to be inside the self-regulatory system and within the
Zone of Proximal Development (ZPD) but outside of the
transference. Many caricatured classical analysts are portrayed
as standing outside of the transference and the ZPD, which
few competent analysts actually do. On the other hand, those
analysts who speak of the co-construction of the transference,

as for example in some contributions from social constructivism to technique, cannot be really analyzing the transference because they are not striving as best as possible to position themselves outside of it (in an absolute sense, obviously an impossibility; in a relative sense, a clinical necessity). Given the inevitability of mutual influences, an analyst can be actively engaged with the patient around many facets of the analysis, yet can strive as best as possible to avoid determining the nature of the transference. The co-construction of the self-regulatory system and the ZPD is an essential prelude to the analysis of the transference and constitutes a major challenge often more prominent during the opening phase of analysis, but a continual challenge throughout the analysis as new conflicts emerge. Although at first glance a contradiction, actually being inside the self-regulatory system or the ZPD and outside the transference is an apt description of a tension that characterizes sound analytic process.

The term *positioning* thus refers to the analyst's relationship to the transference, whether striving to be inside of it (a co-participant) or outside of it (thereby allowing for optimal interpretive leverage). An internalization disorder is not a deficit state but rather is best known and clinically approached as a kind of recruitment of another to accomplish some functional purpose, perhaps somewhat akin to what Anna Freud (1936) intended when she described transference of defense. An internalization disorder is treated by being inside the transference, yet it is not remediated or treated as an absence to be filled in by analytic activity. Nor does an internalization disorder call for excessive kindness or empathic reparenting. Being empathic is a kind of analytic cognition that allows for the unfolding of a process but does not in and of itself provide for remedial experiences. Being kind is a naturally occurring stance of the analyst who is always respectful, always in awe of the complexities of the analysand, but it has no special place in the treatment of an internalization disorder. Being too kind may dangerously habituate an analysand into passive receptivity at an oral level when there might be a yearning for another level.

An internalization disorder, grounded in the angle of attachment, should involve the actions of two persons, both of whom will inevitably behave in ways that the analysand strives to actualize. These actions, although a repetition, are invariably designed to ensure that something that might happen does not. In other words, the processes that prevent internalization reproduce themselves over and over. Although an analysand might say—and believe—that he or she is desperate to effect change, and may pummel the analyst with damning evidence of the analyst's inability to help the analysand effect change, an internalization disorder is actually designed strategically to ensure that no change may take place. Internalization disorders are composed of processes that prevent analysands from taking in what the analyst and the analytic setting are set up to promote. As a consequence of an internalization disorder, the motive not to change competes with both a maturational push to develop and an intrapsychic tendency toward conflict resolution. The possibility of learning is derailed, and cycles of similarity occur. The analysand comes to live in a familiar world that is a kind of unconscious home, and anything that threatens to destroy this psychological home is treated as a danger.

Positioning oneself inside the transference always produces a certain myopia, with a consequent risk of countertransference enactment. It is probably the case that more countertransference actions are induced by the analysand than are described in the analytic literature, which places more emphasis upon countertransference as the product of the analyst's transference to the patient. Under the influence of projection and its externalization kin, we propose that countertransference has a privileged position for the analyst who seeks to move an analysis forward in the face of psychological events that work against progress through the acquisition of insight. This is so because roots of internalization disorders are often not coded symbolically, and so tend to be brought into the analytic transaction as demands on the analyst for work, to pun on Freud's definition of *trieb*. Typically, an analyst will feel the urge to say or do something urgently or an unbidden image will come to mind. When two minds share the same conceptual space, as in an internalization disorder, the concept of projection is altered.

Recruitment is a projection in which separate minds are not assumed or necessarily valued. While some might use the term *projective identification* to depict such an analytic transaction, it can be stressed that the transaction also represents an appeal to perform, as a recruitment gesture to prevent something useful and progressive from happening that might actually be in danger of happening. Recruitment is a demand to replace the potentially unfamiliar with the familiar. When complying with a recruited enactment, what the analyst then enacts is both useful and hurtful; often it is a necessary aspect of the present psychic stability of the analysand, yet at the same time it prevents maturation by keeping the presently existing state of affairs exactly as it is.

Students and supervisees must be wary of the first time in a treatment when they experience an internal demand to do or say something patently obvious to the patient, and wary not to say or do anything, because this countertransference wish can be thought of as a kind of template, whose contour and outline can be studied, like a Rorschach card, in order to learn more about the future of the transference–countertransference matrix. Saying something that is obvious can only render the analysis banal. The analyst who stands outside of the transference is going one way, while the analyst who enters into the flow of regulatory processes is going in the opposite direction. Most analysts used to go in the former direction, while some analysts now go in the latter direction. Few analysts go in both directions, because one's clinical mindset—guided by theory—is not oriented toward two such seemingly contradictory stances.

Transference is something that during an analysis one stands outside of and encourages to flourish. The transference neurosis, when the analyst has gathered its full intensity and in essence focused and concentrated it, drawing it away from the external world, assumes that the analyst is a separate object. Through projection, one's neurosis can become interpretable. This is psychoanalytic bedrock, but the picture now becomes more complicated. What this discussion underscores is that the concept of internalization disorders adds another side to this; the analytic interaction is Janus-faced. One must be both inside the interaction and yet strive to be outside of it. The analyst must be prepared to go in one direction or another at a moment's notice. This is why the distinction between transference and the zone of proximal development is such a crucial one. Paradoxically, first being competent inside the interaction (the ZPD) allows one later to be successful outside of it (i.e., analyzing the transference). The analysis of the transference, in other words, can only take place when the analysand is in the position of being able to internalize its significance. Situating oneself outside of the transference is equivalent to collaborating with, rather than treating, an internalization disorder. This appears to be why Gray (1987), describing the approach of close surface monitoring, makes the following strong statement: "This technical approach aims at providing an opportunity for a maximum of new, conscious ego solutions to conflict, and a minimum of solutions involving new internalizations" (p. 149). He then goes on to explicitly identify his approach as applicable to only those patients who have a "greater capacity for noninternalizing solutions . . .," that is, individuals characterized by compromise formations of conflicts. An internalization disorder is treated by being inside of the interaction as a co-participant, to set into motion processes that will enable the analyst to eventually situate himself outside of the transference, to go about the task of getting it analyzed.

Being "inside" the interaction brings along with it a weighting of the analyst's contribution to the interaction—

and a sense of the need to discuss how one positions oneself with respect to the countertransference. The following clinical vignette will illustrate being inside the interaction:

A 22-year-old man with a history of troubling obsessionality met in four times weekly analysis conducted by a woman analyst. He was quite prone to regression, and as an adolescent this had led to a two-week hospitalization. In the fourth year of the analysis, the pair had been addressing in recent weeks his efforts to make his analyst the mother in charge of everything, so he could passively comply with her dictates and then rage against her swarming control. This allowed him to sequester from the analyst his sexual wishes in the transference. The hour began with his complaint that the room was too cold because the window was open a tad. She felt the temperature was fine. The analyst then remembered that the previous hour had ended with the analysand saying that he felt a fleeting flash of "closeness" that made him so nervous that he had to remain silent for the final fifteen minutes of the hour.

With this in mind, she got up and closed the window, fully aware that she was being drawn into an action from which she would have to emerge and later understand, something to do with again being the hurtful mother-in-charge, of the heat in the room and in the transference. The analyst also was aware that her action felt different, that it might somehow represent a step in analyzing the transference. She had a sudden intrusive image of her own son getting dressed in shorts and a T-shirt that morning, his first day in shorts after a long and cold winter. She then hypothesized that the analysand's enactment was a kind of dressing up, his belated effort to ward off the coldness that had previously served to bury the heat of the sexualization. Once the window was closed, the analysand relaxed in the comfortable feeling of the familiar. He criticized the analyst, recalled the end of the previous hour, and went on to draw the link himself between coldness and closeness. It was not long in the hour before the dreaded issue of his sexuality, heretofore hidden within the labyrinth of obsessional concerns, surfaced in what to him was relative comfort. The analyst let the actions unfold, as she sensed that it had passed beyond a blind and unceasing repetition and was now momentarily serving as a window into the core conflict it defended against. In accord with the principle of multiple function (Waelder 1936), what had at one moment served the aim of repetition was transformed and at another moment served the aim of self-regulation. Later, positioned as best as possible outside of the transference, she analyzed the heretofore dreaded sexual wishes.

This example demonstrates how countertransference can become prioritized as a source of information about the clinical interaction when an internalization disorder is at issue. The analyst participates in the flow of ongoing processes necessary for internalization to optimally occur. In this instance, there is a danger of withdrawing from participation in these processes (e.g., in the service of neutrality, abstinence, or anonymity) and leaving the analysand alone at a time when self-regulation is not smoothly progressing. This is why silence and nonparticipation must be understood as an actual intervention rather than as a nonintervention, because of the dramatic implications it has for self-regulation. The denial of countertransference, thus, is a refusal to face the fact that the analyst contributes to the dynamics of the internalization processes of the analysand.

Often in the case of beginning an analysis, seemingly no amount of insight will help settle a patient into a smooth course of analysis. The analyst comes to recognize the relative intractability of interpreting many issues early in the analysis. Such issues must be lived, at times in an agonized way, to be seen, named, and grappled with. When interpreted early in the analysis, the analysand can only hear criticism, the threat of abandonment, or an accusation, superseding the semantic content of the words comprising the interpretation. Persistent, early interpretation will more than likely drum the analysand out of treatment, a fate Ellman (1991) also cautions against. For these cases in particular (but for all cases in general, too!), how do we bring into motion the concentration of those forces that will enable our verbal interpretations to have the desired impact? How can we promote the analysand's feeling free and able to take in what he or she needs and move along the road to maturation and useful insight? Knowing the key issues implicated is an intellectual counterpart to experiencing them, and the analyst's skill at allowing them to be lived out determines the success or failure of the analysis. This is why Coen (1993) pleads for a "passionate analysis" in which living out these issues in raw form takes center stage in the analysis of individuals who display what Coen terms *pathological dependency*. The analyst must let herself be drawn in, and necessarily so. To stand outside simply leaves the analysand bereft and unable to use the analytic situation. Early in the analysis, if the analyst does not provide for successful self-regulation, the analysis simply cannot continue. Self-regulation refers to being able to survive psychologically on one's own—it is a lived activity and cannot be interpreted into or out of existence.

In addition to interpreting its conflictual elements, the analyst and analysand must live through, together, whatever it is that would enable her to seek a different form of contact. The analyst is drawn into the arena, through her "role responsiveness," of action, of enactment, with which her countertransference emerges as an essential tool for diagnosis. She seeks to position herself inside the disorders of internalization yet outside of the transference. It is often better that the analysand come to understand in her own way her use of contact in the service of attachment rather than the analyst interpreting it because then she would have a different and more efficacious mastery, although this is not always possible. The analyst clears the way for her to approach this understanding. The analyst accepts as inevitable her recruitment as a symbiotic object, and slowly learns what the setting created by the analysand says about what she needs for successful self-regulation and the establishment of a ZPD. This is provided, although titrated. Be clear that titrating does not mean gratifying a wish—it really recognizes the inherent intersubjectivity present in the human condition (as broadly recognized by most current disciplines) and the analytic dyad. Nor does this mean the analyst is removed or uninterpretive—she can be busy interpreting conflicts that are diagnosed within the spectrum of ideational conflict. In this way, the analyst subtly allows for optimal stimulus regulation, but does not exceed what is perceived to be stimulus overload or underload. This is done not because regulating is ipso facto therapeutic, but because it promotes the possibility of inter-nalization through other channels, developments that will circle back and promote later emotional regulation. The ana-lyst only says those words that she can take in, that is, she co-constructs a zone of proximal development and recognizes that actions will often register louder than words; hence, her tact is more important than her interpretive competence. However, more and more as the analysis progresses, she seeks to put action under the hegemony of words when pos-sible. There are many channels of internalization and each case requires time and careful observation before an analyst can diagnose what needs to be handled. The analyst is exquisitely attuned to the fluid, volatile level of organization that is constantly shifting, and whose telltale signs of differ-ence can so easily be overlooked. The analyst's listening ear, the analytic instrument, is always tuned into developmental level in order to infuse content with more structural signifi-cance. For example, the self-lacerating guilt over oedipal sex-ual wishes can be compared to the harsh masochistic trend in separation guilt, or, for that matter, to the self-condemnation of the obsessional fighting against a wish to make things dirty. Most of the clinical manifestations of a disorder of internalization will occur in the lived experience of the here-and-now transference, and through which the analyst can "show" (to use a word frequently employed by the contem-porary neo-Kleinians of London, who I believe are mapping out similar conceptual terrain, although using different underlying assumptions) the analysand issues that are heated up and immediate. Showing presumes that the action has already taken place, whereas interpretation does not make this assumption.

Clearly the countertransference can control the fate of the analysis under these circumstances. It is not so much a clarion cry against unresponsiveness or inexorable neutrality that we are asserting; these are by now tired caricatures, residue of a sad history to which few contemporary structural analysts adhere. One tires of analysts who tritely point to their responsiveness or relative rather than absolute neutrality as signaling that the times have changed. Instead, we are addressing the countertransference stirrings that occur when the analyst "mixes it up" with the analysand around heated and steamy issues. Boesky (1991) makes the intriguing proposal that, unless an analysis in some way gets dragged through the mud of the passions of joint enactment, the analysis will probably fail. Under the conditions of a live and vibrant analysis, countertransference enactment is inevitable. Therefore, it is not the analyst's elimination of the enactments that is called for, but the analyst's skill in monitoring them, recognizing them, and then using self-analysis to take steps backward from them so as to resume an analytic stance and use the issues generated to drive the analysis forward. Unless the countertransference is carefully monitored, and the analyst well analyzed, the analyst will have, at best, only limited leverage to work through and then out of the enactments, so that the transference might eventually become one to a differentiated object in which genital wishes supersede desperate eroticized attachments, structural prerequisites for the pathway to an analysis of the oedipal and other dynamically repressed issues.

References

Arlow, J. (1969). Unconscious fantasy and disturbances of conscious experience. *Psychoanalytic Quarterly* 38:1–27.

Bernstein, R. (1984). *Beyond Objectivism and Relativism*. Philadelphia: University of Pennsylvania Press.

Boesky, D. (1991). The psychoanalytic process and its components. *Psychoanalytic Quarterly* 59:550–584.

Bruner, J. (1964). The course of cognitive growth. *American Psychologist* 19:1–15.

——— (1983). *Child's Talk: Learning to Use Language*. New York: Norton.

Cavell, M. (1989a). Solipsism and community: two concepts of mind in philosophy and psychoanalysis. *Psychoanalytic Contemporary Thought* 12:587–613.

——— (1989b). Interpretation, psychoanalysis, and the philosophy of mind. *Journal of the American Psychoanalytic Association* 37:859–879.

——— (1993). *The Psychoanalytic Mind: From Freud to Philosophy*. Cambridge, MA: Harvard University Press.

Coen, S. (1993). *The Misuse of Persons: Analyzing Pathological Dependency*. Hillsdale, NJ: Analytic Press.

Edelheit, H., reporter. (1968). Language and the development of the ego. *Journal of the American Psychoanalytic Association* 16:113–122.

Ellman, S. (1991). *Freud's Technique Papers*. Northvale, NJ: Jason Aronson.

Emde, R. (1988). Development terminable and interminable, I and II. *International Journal of Psycho-Analysis* 69:23–42, 283–296.

Frank, A. (1969). The unrememberable and the unforgettable: passive primal repression. *Psychoanalytic Study of the Child* 24:59–66. New York: International Universities Press.

Freud, A. (1936). *The Ego and Mechanisms of Defense*. New York: International Universities Press.

——— (1970). The symptomatology of childhood. *Psychoanalytic Study of the Child* 25:19–41. New York: International Universities Press.

Freud, S. (1915). Instincts and their vicissitudes. *Standard Edition* 14:105–141.

Friedman, L. (1978). The barren prospect of a representational world. *Psychoanalytic Quarterly* 39:215–233.

Geertz, C. (1973). *The Interpretation of Cultures*. New York: Basic Books.

Goodman, N. (1984). *Of Mind and Other Matters*. Cambridge, MA: Harvard University Press.

Gray, P. (1987). On the technique of analysis of the superego—an introduction. *Psychoanalytic Quarterly* 61:130–154.

Greenacre, P. (1968). The psychoanalytic process, transference, and acting out. In *The Capacity for Emotional Growth, vol. II*, pp. 762–776. New York: International Universities Press.

Grotstein, J. (1983). A proposed revision of the psychoanalytic concept of primitive mental states: II. The borderline syndrome—section I: Disorders of autistic safety and symbiotic relatedness. *Contemporary Psychoanalysis* 19:570–604.

——— (1989). Self-regulation in schizophrenia. *Psychoanalytic Psychology* 9: 239–268.

Hacking, I. (1983). *Representing and Intervening*. Cambridge: Cambridge University Press.

Hesse, M. (1980). *Revolutions and Reconstructions in the Philosophy of Science.* Bloomington, IN: University of Indiana Press.

Hofer, M. (1984). Relationships as regulators: a psychobiologic perspective on bereavement. *Psychosomatic Medicine* 46:183–197.

Horowitz, M. (1972). Modes of representation of thought. *Journal of the American Psychoanalytic Association* 20:793–819.

Isakower, O. (1939). On the exceptional position of the auditory sphere. *International Journal of Psycho-Analysis* 20:340–348.

Johnson-Laird, P. (1983). *Mental Models.* Cambridge, MA: Harvard University Press.

Kernberg, O. (1976). *Object Relations Theory and Clinical Psychoanalysis.* New York: Jason Aronson.

Khantzian, E. (1978). Ego, self, and opiate addiction: theoretical and therapeutic implications. *International Review of Psycho-Analysis* 5:189–198.

Kitchener, R. (1978). Epigenesis: the role of biological models in developmental psychology. *Human Development* 21:141–160.

Loewald, H. (1960). The therapeutic action of psychoanalysis. *International Journal of Psycho-Analysis* 41:16–33.

Loewenstein, R. (1956). Some remarks on the role of speech in psychoanalytic technique. *International Journal of Psycho-Analysis* 37:460–468.

Luria, A. (1979). *The Making of Mind.* New York: International Universities Press.

Marslen-Wilson, W., ed. (1989). *Linguistic Representations.* Cambridge, MA: MIT University Press.

Mayes, L., and Spence, D. (1994). Therapeutic action in the analytic situation: a second look at the developmental metaphor. *Journal of the American Psychoanalytic Association* 42:789–818.

Pepper, S. (1942). *World Hypotheses*. Berkeley: University of California Press.

Rapaport, D., and Gill, M. (1959). The points of view and assumptions of metapsychology. *International Journal of Psycho-Analysis* 40:153–162.

Rommetveit, R. (1974). *On Message Structure: A Framework for the Study of Language and Communication*. London: Wiley.

Rorty, R. (1979). *Philosophy and the Mirror of Nature*. Princeton, NJ: Princeton University Press.

Schafer, R. (1968). *Aspects of Internalization*. New York: International Universities Press.

Shapiro, T. (1979). *Clinical Psycholinguistics*. New York: Plenum.

Stern, D. (1985). *The Interpersonal World of the Infant: A View from Psychoanalysis and Developmental Psychology*. New York: Basic Books.

Stolorow, R., Brandchaft, B., and Atwood, G. (1987). *Psychoanalytic Treatment: An Intersubjective Approach*. Hillsdale, NJ: Analytic Press.

Trevarthen, C. (1979). Communication and cooperation in early infancy: a description of primary intersubjectivity. In *Before Speech: The Beginning of Interpersonal Communication*, ed. M. M. Bullowa. New York: Cambridge University Press.

Vygotsky, L. (1988a). Thinking and speaking. In *The Collected Papers of L. S. Vygotsky*, vol. 1, ed. R. W. Rieber and A. S. Carton, pp. 39–288. New York: Plenum.

——— (1988b). Emotions and their development in childhood. In *The Collected Papers of L. S. Vygotsky*, vol. 1, ed. R. W. Rieber and A. S. Carton, pp. 325–338. New York: Plenum.

——— (1988c). Imagination and its development in childhood. In

The Collected Papers of L. S. Vygotsky, vol. 1, ed. R. W. Rieber and A. S. Carton, pp. 339–350. New York: Plenum.

Waelder, R. (1936). The principle of multiple function. *Psychoanalytic Quarterly* 5:45–62.

———— (1962). Psychoanalysis, scientific method, and philosophy. In *Psychoanalysis: Observation, Theory, Application*, pp. 248–274. New York: International Universities Press, 1976.

Wertsch, J. (1985). *Vygotsky and the Social Formation of Mind.* Cambridge, MA: Harvard University Press.

———— (1990). *Voices of the Mind.* Cambridge, MA: Harvard University Press.

Wilson, A. (1989). Levels of adaptation and narcissistic psychopathology. *Psychiatry* 52:218–236.

Wilson, A., and Passik, S. (1993). Explorations in presubjectivity. In *Hierarchical Concepts in Psychoanalysis: Research, Theory, and Clinical Practice*, ed. A. Wilson, and J. Gedo, pp. 76–126. New York: Guilford.

Wilson, A., Passik, S., and Faude, J. (1990). Self-regulation and its failures. In *Empirical Studies in Psychoanalytic Theory*, vol. III, ed. J. Masling, pp. 149–211. Hillsdale, NJ: Lawrence Erlbaum.

Wilson, A., and Weinstein, L. (1992a). An investigation into some implications of a Vygotskian perspective on the origins of mind: psychoanalysis and Vygotskian psychology, part I. *Journal of the American Psychoanalytic Association* 40:357–387.

———— (1992b). Language and clinical process: psychoanalysis and Vygotskian psychology, part II. *Journal of the American Psychoanalytic Association* 40:725–759.

———— (1996). The transference and the zone of proximal development *Journal of the American Psychoanalytic Association* 44:167–200.

Wittgenstein, L. (1953). *Philosophical Investigations.* London: Blackwell.

Zinchenko, V. (1985). Vygotsky's ideas about units for the analysis of the mind. In *Culture, Communication, and Cognition: Vygotsky Perspectives,* ed. J. Wertsch, pp. 94–118. New York: Cambridge University Press.

7

The Analyst's Role in Internalization

Andrew B. Druck

How do our perspectives on early development affect our clinical approach? I will focus on the individual's psychological development from a part-object state, where others are needed to help the developing psychological structure maintain various psychological functions, to a whole-object state, where the individual has gained the capacity to perform such functions for him- or herself. Part and whole object relations, in this sense, refer both to different kinds of transferences and to different levels of psychic structural development that underlie these modes of object relationship.

It is widely assumed that adequate ego and superego development requires optimal internalization of the parent. Internalization is the means towards an end; it is the developmental path through which whole-object development and its structural consequences are achieved. Wilson and Prillaman state in Chapter 6 that internalization is an "aspect of what psychoanalysts treat" (p. 190). Here they condense a complex matter: difficulties in optimal internalization as they affect the patient's level of functioning are one aspect of what analysts treat, but internalization of the analyst is also one important aspect of what makes for a successful treatment. How we work with patients where an important prognostic and mutative factor in treatment—the patient's capacity to internalize the analyst and his analytic functioning—is itself problematic becomes a central question.

Traditional psychoanalytic theory has assumed whole-object relations and has focused on pathology based on conflict embedded within these developed structures. In this situation, the analyst's optimal role has been to interpret to the patient the manner in which this unconscious conflict becomes present in the transference. The analyst does so from what Wilson and Prillaman term a position "outside the transference." The analyst proceeds in this way because our assumption of whole-object functioning presumes that the analyst is not needed for structural maintenance. Thus focus only needs to be on the transference.

Once we find part-object functioning and its associated structural difficulties, we assume much closer interpenetration between a patient's dynamic conflicts, defenses, and level of structural functioning. The patient cannot contain conflict at his or her level of structure and regresses to an earlier level of structural functioning. Such regression is clinically demonstrable, as Pine (1974) has discussed in his classic paper on pathology of the differentiated versus the undifferentiated other, and as Daniel has demonstrated in Chapter 5 of this volume. We also find a closer connection between patient and analyst, both dynamically, through projective identification, and structurally, when the patient needs the therapeutic object relationship as an aid in structural regulation. For both these reasons, the boundary between patient and analyst becomes much more porous than with the neurotic patient. This is why the analyst's subjective reactions to his or her patient become a major area of focus, as we see in the chapters under discussion. Part-object functioning and its structural consequences thus challenge our traditional conception of the optimal analytic stance and of what is mutative.

Analysts disagree on how our conception of the analyst's

role changes because they emphasize different parts of the problem in patients with structural difficulty (see Chapters 4, 5, and 6). In brief, Fosshage and Lichtenberg, in Chapter 4, emphasize the patient's need of the analyst for the development and consolidation of a positive sense of self. They emphasize a combination of close attunement to the patient that facilitates selfobject experience, along with analysis of problematic organizing patterns. This combination facilitates psychological reorganization, self-development, and the capacity for new and broadened relational perspectives. Daniel stresses in her chapter the severe and early conflicts and defenses particular to the paranoid-schizoid and depressive positions rather than the patient's needs for help in self-regulation. While Fosshage and Lichtenberg in Chapter 4 concentrate on the patient's attempts at adaptation and attachment, Daniel focuses on the patient's attempts at defense. Daniel believes that the analyst must interpret unconscious dynamic conflicts that interfere with experiencing the analyst as a good object. As these conflicts are interpreted, the patient is able to experience the analyst as a good object, internalize her, and move from the paranoid-schizoid to the depressive position. Wilson and Prillaman discuss in their chapter the manner in which early adaptational needs become intertwined with dynamic conflict. For Wilson and Prillaman, the analyst simultaneously functions as a symbiotic object and as a transference object. They attempt to conceptualize the analyst's function as both a structural aid and an interpreter of transference stemming from unconscious conflict.

Fosshage and Lichtenberg (Chapter 4) posit an "organization model" of transference that focuses on the manner in which we constantly organize our experiences according to unconscious themes. Fosshage (1994) says that

> we are always constructing reality every bit as much as we are perceiving it. The predominant ways in which we have come to see ourselves and ourselves in relation to others are the affect-laden thematic organisations that variably shape our experience. These *affect-laden organising principles or schemas* . . . do not constitute a supposed "objective reality" but are always contributing to the construction of a subjectively-experienced "reality." [p. 267]

Both patient and analyst contribute to this organization of reality. Transference is the patient's continuing assimilation of the present analytic relationship along the lines of his personal thematic structures.

In this volume, Fosshage and Lichtenberg emphasize that transference should not be seen as displacement, and they believe that "the organizing activity we speak of does not produce replicas of prior experiences, but creates new lived experiences modeled on significant features of prior experiences" (p. 133). Fosshage (1994) states that "schemas are activated, not transferred" (p. 271). In their clinical example of a female patient, Fosshage and Lichtenberg emphasize that the patient's disappointment in the priest, in the analyst, and in her parents is not displacement but reaction that is "organized by the same thematic emotional experience that was triggered in her perception of each situation" (see p. 128).

Fosshage and Lichtenberg state that patients approach the analyst with a combination of hope for what they term a

needed selfobject experience and dread of repeating past relationships. The latter expectation is closest to the Freudian
model that sees transference as stemming from conflict based
on earlier experiences. Conflict arises as patients are caught
between their hope that they will receive a desired selfobject
experience from the analyst and fear that expectations from
aversive interpersonal experiences will be repeated. As in the
clinical example in Chapter 4, for example, the patient is conflicted between trusting the analyst, who could be helpful,
and fearing the analyst, who could be sexually threatening or
abandoning. The analytic work is focused on a combination
of providing needed vitalizing selfobject experiences while
also illuminating problematic and conflicted organizing patterns and expectations, all within a transferential field that is
co-constructed, to a greater or lesser degree, by the patient
and analyst. Fosshage and Lichtenberg devote much of their
chapter to illustrating how motivational theory expands our
understanding of the various kinds of selfobject experiences
that emerge in the analytic relationship.

Fosshage and Lichtenberg's understanding of transference (an "organization" model [Fosshage 1994]) differs
from the traditional Freudian model (what Fosshage [1994]
terms a "displacement" model) in many ways. Two important points that are relevant to the topic in question need
emphasis. First, while Fosshage and Lichtenberg do see what
they term *illumination of conflict* as significant, the conflictual situations they discuss tend to be relational ones—
focused on various combinations of fears and hopes regarding object relationships. They do not consider conflict as aris-

ing from internally generated wishes that lead to anxiety and various forms of defense. This, of course, is more characteristic of Freudian and Kleinian views of transference.

Second, while this understanding of selfobject experience may be common among self psychologists, it is new to my Freudian ear. Kohut's profound contribution to our field was his identification of a clinically demonstrable form of part-object transference and development of a theory that helped us understand it. Kohut (1971) originally defined a selfobject as narcissistically cathected, and he distinguished it from what he termed "true objects . . . which are cathected with object-instinctual investments, i.e., objects loved and hated by a psyche that has separated itself from the archaic objects, has acquired autonomous structures, has accepted the independent motivations and responses of others, and has grasped the notion of mutuality" (p. 51).

In his later work (1984), Kohut emphasized the constant intertwining of selfobjects and true objects. Shane and Shane (1993), in their review of developments in self psychology since Kohut's death, discuss differences between self psychologists in just this area. They write as follows: "Basch stands alone among those who use the selfobject concept in viewing the requirement for selfobject function as indicative of an *endangered* or *fragmented* self; the others conceive of the need for a selfobject relationship as part of the *normal* self functioning" (p. 787).

The first way of thinking about selfobject experience is consistent with the work of contemporary Freudian analysts such as Wilson and Prillaman in their chapter, and analysts such as Bach (1983, 1994), Grunes (1984), Adler (1989, Buie and Adler 1973), and perhaps Loewald (1980). They would see these selfobject experiences as stemming from the adaptive and structurally aiding role of part-object relationships. However, when Fosshage and Lichtenberg speak of selfobject experiences in their chapter, they are not discussing part-object relations at all. Selfobject experiences are personally encountered events that may occur within *any kind* of object relationship, whether they are whole or part-object relationships. Thus Fosshage and Lichtenberg define selfobject experiences as "experiences that contribute to vitality" (p. 134). Selfobject experiences are one dimension in the complexity of human functioning. There are many kinds of selfobject experiences within the differing motivational themes. They contrast these selfobject experiences, which "are particularly influential organizers of new creations" and "facilitate the psychological development of the child" (p. 134) with those experiences that involve thematic misattunements [which] . . . will tend to organize expectations that lead to the dominance of one motivational system, often aversiveness, at the expense of other motivations, such as attachments" (p. 134).

How can we tell whether an experience is a selfobject experience? Certainly we might say that positive experiences are selfobject experiences while negative experiences are not. However, are *all* positive experiences selfobject experiences? How do we know if an experience is enjoyable or vitalizing, or is every enjoyable experience presumed to be vitalizing? Further, what about negative experiences that, because they touch on certain conflicted but affectively important internalized objects, are hurtful and also quite vitalizing? Answering these questions not only would help us understand the concept better but would help us understand, from a self psychological point of view, analytic boundary issues such as analytic abstinence and neutrality. For example, perhaps the same tension Wilson and Prillaman discuss in Chapter 6 between wanting to support a weakened ego and yet interpret a transference state might be operative in a self psychologist who would want to facilitate selfobject experiences but not participate in experiences that might be pleasurable but not vitalizing.

In Chapter 5 Daniel discusses the role of internalization, which she terms *introjective identification*, in strengthening the ego and helping the patient move from the paranoid-schizoid position to the more developmentally advanced depressive position. Daniel believes that positive internalization is not only a matter of the object's attunement to the patient. Rather, the patient is conflicted about introjective identification.

Implicit in the concept of Kleinian positions is Loewald's (1980) idea that patients at different levels of psychological structure organize reality differently. Daniel illustrates the role of internal conflict in transforming one's experience of the object and of reality itself. For her, the "dynamic primitive inner world becomes externalized through the analytic process" (p. 185) and in the transference.

Thus Daniel discusses her patient's experience of hearing the sounds of motor bikes revving outside. At that time, Daniel notes that the patient was feeling relatively integrated within herself, and in good contact with the analyst. She also had more awareness, at that moment, of external reality, and so she was more able to tolerate the intrusion and the irritation provoked by it. . . . On other occasions when she had felt more disturbed, precarious, or competitive, and outside noises intruded, she had gone into grinding resentment or paroxysms of rage, declaiming that I had failed to protect her or that I had even deliberately engineered the interference. When functioning at a paranoid-schizoid level, she would have felt the bikes and I were revving up against her, to get her out: both parents united against her infantile self. Alternatively, she might have projected onto the revving bikes her own hatred of the analyst who "kicked her out." Either way, she would have felt terribly persecuted. [p. 181–182]

The madness, so to speak, begins in the patient and is projected into the analyst, who is seen as a part object. The analyst first becomes transformed by the madness and must then contain and interpret what has been projected back into the patient.

Daniel never forgets her primary focus: the patient's internal world and internal object relationships. She ends her clinical vignette not with a statement about the patient's relation to her analyst, but with the following quote from the patient: "When I can't reach good relationships inside me, I can't write," which Daniel explains as follows: "She had become able to know her own psychic state" (p. 180).

Daniel believes that appropriate interpretations of unconscious conflict help the patient internalize the analyst as a good object and move up the developmental ladder, from the paranoid-schizoid to the depressive position. She does not address the question of the patient's capacity to hear, tolerate, and utilize interpretations that are usually at variance with the patient's conscious experience. One question to consider is how Daniel understands why the patient would take her in and consider her interpretations, particularly in periods of extreme regression. While Daniel's discussion of dynamic factors interfering with internalization is subtle and developed, it would be illuminating to hear more discussion from her regarding patients' motives *for* change and ego capacities that facilitate such change.

Interestingly, for Fosshage and Lichtenberg, the kinds of interpretations suggested by Daniel are precisely the wrong thing to do because they will make the patient feel misunderstood and will threaten the patient's psychic equilibrium, which depends on the analyst's allowing himself to be used as a selfobject. It would actually be quite illuminating to ask all the participants the following question, which is as close as we can get to an empirical test: Let us assume that these clinicians obtained excellent results with their patients. How, then, would they explain each other's success, given that they followed almost opposing technical recommendations?

Wilson and Prillaman believe that the connection between patient and analyst facilitates the patient's capacity and willingness for the analytic work. This connection is not positive transference, where the patient works because his motive is to be a good patient and please the analyst. Instead, they refer to the way that the patient, at any given background moment in treatment, internalizes the analyst on one level and, through this moment-by-moment internalization, gains the capacity for a higher level of structural capability—the analyst is simultaneously an adaptive part-object and a whole transference object in a manner that constantly shifts. The former aspect of the analyst's relationship supports the patient's capacity to analyze the latter. Wilson and Prillaman present a technical guideline that follows from this theoretical assumption: "an analyst strives to be inside the self-regulatory system and within the Zone of Proximal Development (ZPD) but outside of the transference" (p. 216).

Wilson and Prillaman emphasize the interconnection of structural needs and dynamic conflict. What some would term structural deficits they term *disorders of internalization.* They believe that "an internalization disorder is not a deficit state but rather is best known and clinically approached as a kind of recruitment of another to accomplish some functional purpose, perhaps somewhat akin to what Anna Freud intended when she described transference of defense" (p. 217). They further believe that these disorders are composed of processes that prevent analysands from taking in what the analyst and the analytic setting are set up to promote.

The category "disorders of internalization" that Wilson and Prillaman use is quite broad and could include patients at varying levels of psychic structure. However, there are different kinds of disorders of internalization. Many of these involve a substantial degree of structural deficit. I am not clear why they call these "transference states" rather than deficit states. By "deficit state" I mean clinical phenomena that can be clearly demonstrated as showing some degree of structural weakness and that we may presume to stem from a combination of dynamic conflict within the context of an undeveloped psychic structure. Obviously there are degrees to the severity of the particular structural weakness. If these are predominantly transference states and not deficit states, then Wilson and Prillaman's entire therapeutic rationale comes into question. Why not simply interpret these states, as Anna Freud advocates? Perhaps they mistakenly assume that to call something a deficit state means that it is something to be "remediated" or "treated as an absence to be filled in by analytic activity" (p. 217). However, their subtle and complex suggestions regarding the analyst's stance in responding to the interweaving of these states with conflictual material shows that they must realize that one can acknowledge a deficit without abandoning an analytic focus. I agree with them on all of their technical recommendations, although we would label the phenomena under discussion differently.

Wilson and Prillaman also consider disorders of internalization as operating to prevent the analysand from taking in what the analyst has to offer. Sometimes they discuss these disorders as a motive not to change while at other times they seem to be speaking of the patient's inability to regulate himself. These are different things. Perhaps they mean that when the analyst acts to help the patient with structural stability, what they term an *enactment*, then "what the analyst then enacts is both useful and hurtful; often it is a necessary aspect of the present psychic stability of the analysand, yet at the same time prevents maturation by keeping the presently existing state of affairs exactly as it is" (p. 220). However,

Wilson and Prillaman later discuss how the analyst attempts to maintain a gap between what the patient needs and what he can do for himself and they do say that meeting certain needs helps the patient work with other aspects of the transference.

It is crucial to clinically identify states that have to do with structural need and differentiate them from early transference wishes and resistances. With regard to structural needs, or deficits, some involve movement toward the analyst rather than processes destructive to treatment. For example, disorders of self-esteem regulation, libidinal object constancy, or a tendency toward panic anxiety and emotional flooding often involve desperate and conflicted attempts by the patient to get the inner controls that he needs from the analyst. This would also apply to what Wilson and Prillaman term *disorders of self regulation*. Other disorders of internalization, however, particularly the kinds of superego difficulties described by Joseph (1982) and others, involve movement away from the analyst and processes that are destructive to the analysis. (A more extensive discussion of this issue will be found in Druck [in press].)

Despite theoretical quibbles with Wilson and Prillaman, their attempt to examine how structural and dynamic processes interact is typical of the best writing in contemporary Freudian psychoanalysis. They seem to combine the best of the Fosshage and Lichtenberg chapter and the Daniel chapter. Like Daniel, they are alert to the role of internally generated unconscious conflict in structuring the psychoanalytic situation, and, like Daniel, they are fully aware of the way in which conflict can lead to regression to a lower level of structure as well as to anxiety and defense within the same level of structure. Daniel, Wilson, and Prillaman are alert to different kinds of conflict at different levels of structure. Like Fosshage and Lichtenberg, Wilson and Prillaman are aware of the analyst's role as an aid to develop structure for certain patients and the crucial role of this support in helping the patient do the analytic work.

Much of our psychoanalytic history deals with identification of different kinds of dynamic conflict and their effect on symptom formation and the transference. A major challenge for psychoanalysis today is to identify clinical phenomena that demonstrate different levels of structural functioning. We must then show how the dynamic conflicts we all know so well interact at different levels of psychic structure. This means that we must move to an ever more complex way of viewing subtle interacting levels of functioning and toward looking more carefully at the many simultaneously functioning aspects of the analyst's role.

References

Adler, G. (1989). Uses and limitations of Kohut's self-psychology in the treatment of borderline patients. *Jounal of the American Psychoanalytic Association* 37:761–785.

Bach, S. (1983). *Narcissistic States and the Therapeutic Process.* New York: Jason Aronson.

——— (1994). *The Language of Perversion and the Language of Love.* Northvale, NJ: Jason Aronson.

Buie, D. H., and Adler, G. (1973). Definitive treatment of the borderline personality. In *International Journal of Psychoanalytic Psychotherapy*, vol. 9, 1982–1983, ed. R. Langs, pp. 51–87. New York: Jason Aronson.

Druck, A. B. (in press). Deficit and conflict: an attempt at integration. In *Contemporary Freudian Technique: Legacy, Controversy, and Critical Issues*, ed. C. Ellman, S. Grand, M. Silvan, and S. J. Ellman. Northvale, NJ: Jason Aronson.

Fosshage, J. (1994). Toward reconceptualizing transference: theoretical and clinical considerations. *International Journal of Psycho-Analysis* 75(2):265–280.

Grunes, M. (1984). The therapeutic object relationship. *Psychoanalytic Review* 71:123–143.

Joseph, B. (1982). Addiction to near death. In *Psychic Equilibrium and Psychic Change*, ed. M. Feldman and E. B. Spillius, pp. 27–38. London: Routledge.

Kohut, H. (1971). *The Analysis of the Self*. New York: International Universities Press.

——— (1984). *How Does Analysis Cure?* Chicago: University of Chicago Press.

Loewald, H., ed. (1980). Ego and reality. In *Papers on Psychoanalysis*. New Haven: Yale University Press.

Pine, F. (1974). Pathology of the separation-individuation process as manifested in later clinical work. *International Journal of Psycho-Analysis* 60:225–242.

Shane, M., and Shane, E. (1993). Self psychology after Kohut: one theory or many. *Journal of the American Psychoanalytic Association*, 41(3):777–797.

Credits

The editors gratefully acknowledge permission to reprint material from the following sources:

"The Contribution of Self- and Mutual Regulation to Therapeutic Action: A Case Illustration," by Frank M. Lachmann and Beatrice Beebe, in *Basic Ideas Reconsidered: Progress in Self Psychology*, volume 12, ed. A. Goldberg. Copyright © 1996 by The Analytic Press. Reprinted by permission of the authors, the editor, and The Analytic Press.

"Forging a Link between Basic and Clinical Research: Developing Brain," by W. A. Himwich, in *Biological Psychiatry*, volume 10. Copyright © 1975 by Plenum Press. Used by permission of Plenum Press.

"Mother–Infant Mutual Influence and Precursors of Psychic Structure," by B. Beebe and F. M. Lachmann, in *Progress in Self Psychology*, volume 3, ed. A. Goldberg. Copyright © 1988 by The Analytic Press. Used by permission of The Analytic Press.

Emotion and Early Interaction, by T. Field and A. Fogel. Copyright © 1982 by Lawrence Erlbaum Associates. Used by permission of Lawrence Erlbaum Associates.

The Perceived Self, edited by U. Neisser. Copyright © 1993 by Cambridge University Press. Used by permission of Cambridge University Press.

Index